Great Explorations in Math and Science (GEMS)

Lawrence Hall of Science,
University of California, Berkeley

SPACE SCIENCE SEQUENCE FOR GRADES 3–5

Unit 2 Earth's Shape and Gravity

The Space Science Sequence is a collaboration between the
Great Explorations in Math and Science (GEMS) Program
of the Lawrence Hall of Science,
University of California at Berkeley and the
NASA Sun–Earth Connection Education Forum
NASA Kepler Mission Education and Public Outreach
NASA Origins Education Forum/Hubble Space Telescope
NASA Solar System Education Forum
NASA IBEX Mission Education and Public Outreach
Special advisors: Cary Sneider and Timothy Slater
Foreword by Andrew Fraknoi

s and Space Administration

GEMS Space Science Sequence was provided by the
orums and Missions listed on the title page.

Great Explorations in Math and Science (GEMS) is an ongoing curriculum development program and growing professional development network. There are more than 70 teacher's guides and handbooks in the GEMS Series, with materials kits available from Carolina Biological. GEMS is a program of the Lawrence Hall of Science, the public science education center of the University of California at Berkeley.

Lawrence Hall of Science
University of California
Berkeley, CA 94720-5200
Director: Elizabeth K. Stage

Project Coordinator: Carolyn Willard
Lead Developers: Kevin Beals, Carolyn Willard
Development Team: Jacqueline Barber, Lauren Brodsky, John Erickson, Alan Gould, Greg Schultz
Principal Editor: Lincoln Bergman
Production Manager: Steven Dunphy
Student Readings: Kevin Beals, Ashley Chase
Assessment Development: Kristin Nagy Catz
Evaluation: Kristin Nagy Catz, Ann Barter
Technology Development: Alana Chan, Nicole Medina, Glenn Motowidlak, Darrell Porcello, Roger Vang

Cover Design: Sherry McAdams, Carolina Biological Supply Co.
Internal Design: Lisa Klofkorn, Carol Bevilaqua, Sarah Kessler
Illustrations: Lisa Haderlie Baker
Copy Editor: Kathy Kaiser

This book is part of the *GEMS Space Science Sequence for Grades 3–5*.
The sequence is printed in five volumes with the following titles:
Introduction, Science Background, Assessment Scoring Guides
Unit 1: *How Big and How Far?*
Unit 2: *Earth's Shape and Gravity*
Unit 3: *How Does the Earth Move?*
Unit 4: *Moon Phases and Eclipses*

Published by Carolina Biological Supply Company. 2700 York Road, Burlington, NC 27215.
Call toll-free 1-800-334-5551.

Printed on recycled paper with soy-based inks.

ISBN 978-0-89278-334-2

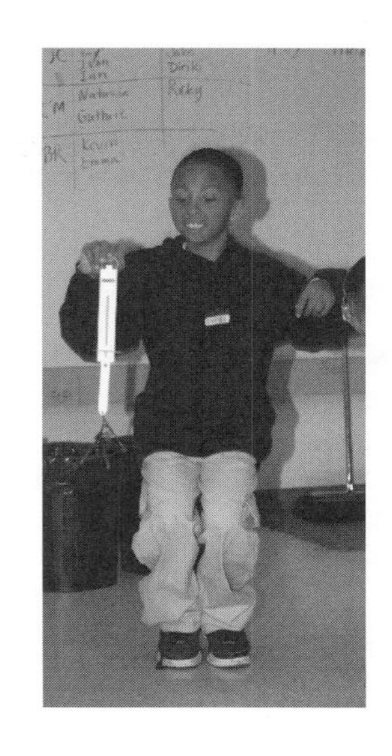

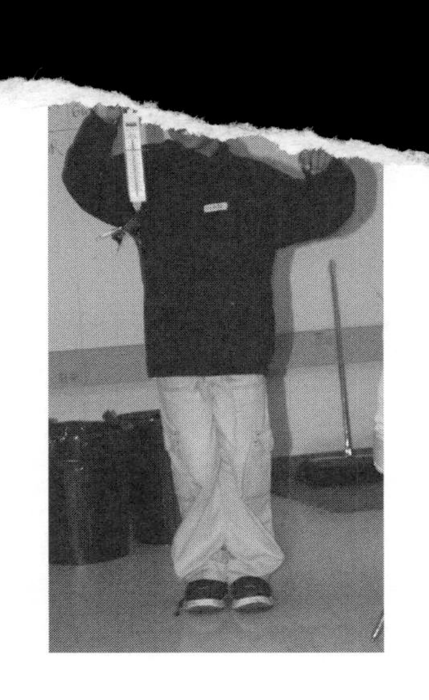

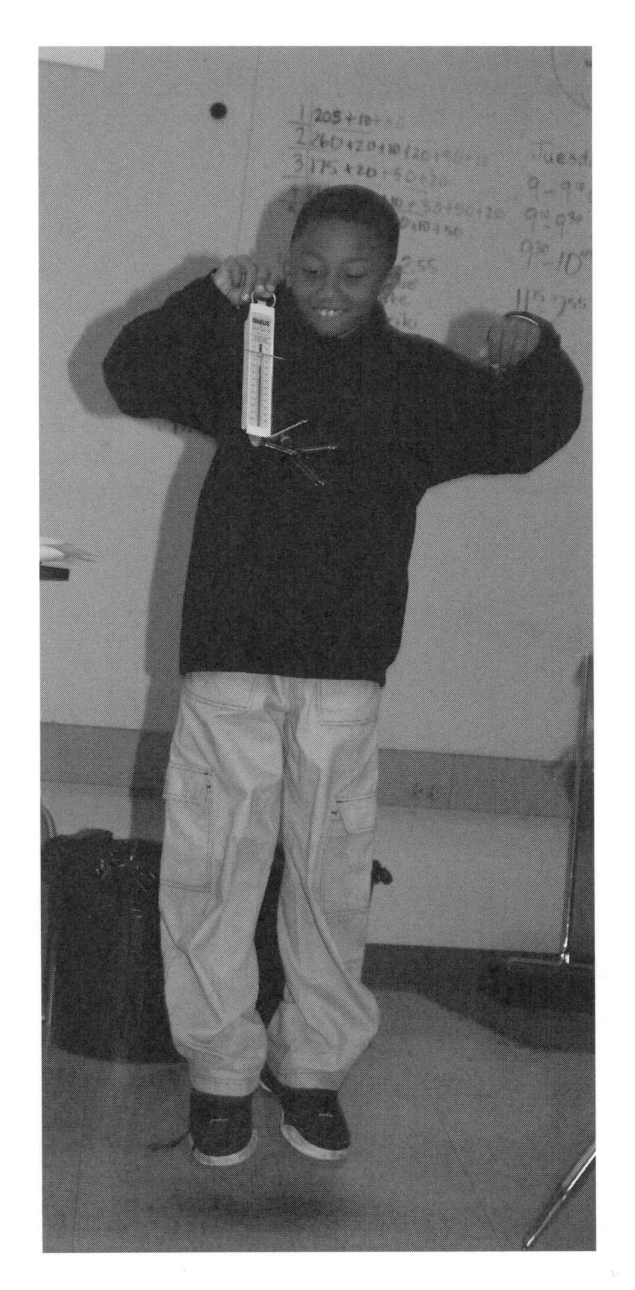

UNIT 2

EARTH'S SHAPE AND GRAVITY

SESSION SUMMARIES (6 Sessions)

2.1: Ideas About Earth's Shape and Gravity

A questionnaire launches your students into animated discussions about the implications of the ball-shaped Earth, giving them the opportunity to examine their preconceptions and begin to explore the Earth's shape and gravity. First, each student fills out the *Pre–Unit 2 Questionnaire* individually. Then, the questionnaire is used again, as students work in small groups to discuss their ideas about each of the questions. They use a globe to explain their ideas about the Earth's shape and gravity. Finally, you facilitate a class discussion of the questions.

2.2: What Shape Is the Earth?

The session begins with an introduction to key concepts about models. In a large–group discussion, students examine two different models of the Earth: a spherical model and a flat-Earth model. Students learn that the spherical Earth model is based on evidence. After defining how scientists use evidence, the class embarks on a virtual orbit around the Earth through a series of satellite images. They observe that the Earth looks flat from close up but spherical from farther away. After their orbital tour, the class reviews two questions about the Earth's shape from the *Pre–Unit 2 Questionnaire.* Three key concepts about Earth's shape are posted on the class concept wall. The students then work in small groups called evidence circles to respond to four statements by (fictional) people who believe that the Earth is flat. Students get practice in using evidence-based arguments, and in the process, they may solidify their own and their fellow students' conceptions about the Earth.

2.3: Gravity

The session begins with a quick demonstration and an introduction to some basic concepts about gravity. Students then look at photographic evidence of how gravity affects people on different parts of the Earth. With this evidence in mind, students revisit question #4 from the Unit 2 questionnaire, which asks what direction rocks would fall on various parts of the Earth. The concept that the gravitational pull between the Earth and the rocks pulls all the rocks toward the center of the Earth leads to a discussion of how we can "beat" gravity. In small–group evidence circles, students discuss the effects of gravity by answering the question, "Is gravity strong or weak?" They share their evidence back in the large group. Students then read about Yuri Gagarin, the first person ever to orbit the Earth in space. They learn that although he felt weightless in his spaceship while in orbit, Gagarin did not escape the pull of gravity between himself and the Earth.

2.4: Weightlessness

The session begins with an introduction to a spring scale as a device used to measure the gravitational pull between an object and the Earth. Student pairs rotate through a series of learning stations featuring drawings of people in different situations and discuss whether they think a person might feel weightless there, and whether or not gravity is present. In a follow-up class discussion, students learn that although a person might feel weightless in every one of the situations, gravity is present in every situation. Next, the class reads *The Vomit Comet*. The reading explains the causes of weightlessness and reinforces the concept that gravity exists everywhere in the universe. In a follow-up demonstration, a student weighs an object with a spring scale. The class carefully observes and records what happens to the object's weight as the student jumps and lands with the spring scale in hand. They discuss how the spring scale shows temporary weightlessness, even though the gravitational pull between the object and the Earth remains in effect. Finally, the class enjoys footage of people experiencing weightlessness in the Vomit Comet and in spacecraft.

2.5: Gravity and Air

The teacher drops a binder clip in front of the class, and students are introduced to some ideas about how fast objects fall on Earth. Pairs of students are given materials, and are challenged to find a way to get one binder clip to fall more slowly than another. They demonstrate their inventions for the class, and discuss how air resistance affects falling objects. Next, the teacher simultaneously drops a feather and a hammer as a demonstration. Students discuss why one dropped more slowly than the other. They predict what would happen if the same test were performed on the Moon. They watch a brief film clip of an astronaut performing this test on the Moon, and see the feather plummet to the ground just as fast as the hammer does. A closing discussion confirms that the Moon does have gravity, but no air.

2.6: Gravity Beyond Earth

The final session in the unit begins with a simulated mission to the Moon, as students are shown a series of 22 images from the Apollo 11 mission. After viewing the mission, students review the concept of gravity by discussing two questions in evidence circles, and backing up their answers with evidence from the Apollo 11 images: "Is there gravity on the Moon?" "Is there air on the Moon?" The evidence that there is indeed gravity on the Moon, but no air, is shared in a class discussion. Students deconstruct the common misconception that air is what keeps objects on the ground. The Apollo 11 images also serve to reinforce the concept of the spherical shape of the Earth and Moon. To conclude Unit 2, students take the *Post–Unit 2 Questionnaire* to find out how their ideas have changed since the beginning of the unit.

SESSION 2.1

Ideas About Earth's Shape and Gravity

Overview

A pre-unit questionnaire launches your students into animated discussions about the implications of a ball-shaped Earth, giving them the opportunity to examine their preconceptions and to explore the Earth's shape and gravity. First, each student fills out the questionnaire individually. Then, students work in small groups to discuss their ideas about each of the questions. They use a globe to explain their ideas about the Earth's shape and gravity.

Finally, you facilitate a class discussion of the questions, keeping in mind that ideas and insights about the Earth's shape and gravity develop gradually. Getting the "right answer" quickly is not as important as the critical thinking skills that students develop as they struggle to apply their mental models of the Earth to both real and imaginary situations. To give students time to reflect, we recommend that you refrain from giving them the correct responses to the questions until the subsequent sessions in Unit 2, after they have had opportunities to test, discuss, and debate their ideas.

Ideas About Earth's Shape and Gravity	Estimated Time
Taking the *Pre-Unit 2 Questionnaire*	20 minutes
Conducting small-group discussions	20 minutes
Conducting a class discussion	20 minutes
TOTAL	**60 minutes**

What You Need

For the class
- ❑ chalkboard, chart paper, or overhead projector

For each group of three or four students
- ❑ 1 copy of the *Pre–Unit 2 Questionnaire*, from the student sheet packet
- ❑ 1 Earth globe or other large ball (globes are better than balls, and transparent, inflatable globes are even better than opaque globes)
- ❑ 1 bowl or roll of tape (used as a base to support globe)

For each student
- ❑ 1 copy of the *Pre–Unit 2 Questionnaire* from the student sheet packet
- ❑ 1 pencil

Unit Goals

Earth, like all planets and stars, is round like a sphere.

Earth's gravity pulls things toward the center of Earth.

Gravity is everywhere.

TEACHER CONSIDERATIONS

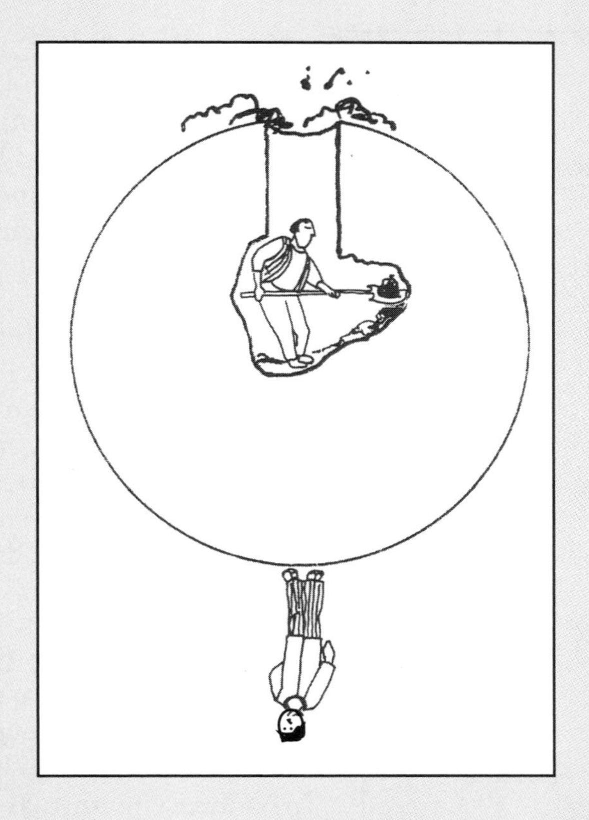

TEACHING NOTES

Misconceptions about the Earth's shape: Despite the evidence of our senses, we are told as early as the first and second grades that the Earth is really shaped like a ball, that the Earth is round. We hear that you could "dig a hole all the way to China," or that people in faraway lands live "down under your feet." Young children may construct private models to reconcile what they are told about the round Earth with the evidence of their eyes. For example, because the Earth looks flat, they may envision the Earth as "round," but like a plate.

When students reach upper elementary grades, they are capable of understanding the Earth's spherical shape and related concepts about gravity. However, to prepare them to absorb these concepts, allow students to air and reexamine the mental models constructed in prior years.

See page 31-33 in the *Background* section for other common misconceptions about Earth's shape and gravity.

Key Vocabulary

Science and Inquiry Vocabulary

Evidence

Scientific Explanation

Model

Scale Model

Prediction

Scientist

Three–Dimensional (3-D)

Two–Dimensional (2-D)

Space Science Vocabulary

Atmosphere

Air Resistance

Force

Mass

Gravity

Weightless

Satellite

Orbit

Diameter

Sphere

System

PRE–UNIT 2 QUESTIONNAIRE, PAGE 1

Session 2.1 Student Sheet Name ____________________

Pre–Unit 2 Questionnaire, Page 1

1. Why is the Earth flat in picture #1 and round in picture #2? (Circle the letter of the best answer.)

A. They are different Earths.
B. The Earth is round like a ball, but people live on the flat part in the middle.
C. The Earth is round like a ball, but it has flat spots on it.
D. The Earth is round like a ball, but it looks flat because we see only a small part of the ball.
E. The Earth is round like a plate, so it seems round when you are over it and flat when you are on it.

2. Which are true statements? Circle all that are true.

A. The Moon has no gravity because it has no air.
B. Gravity is an invisible force.
C. There is no gravity in space.
D. There is a pull of gravity between all objects.
E. If there were no air on Earth, people would float out into space.
F. Gravity keeps the Moon in Earth's orbit.

3. Pretend that the Earth is glass and you can look straight through it. Which way would you look, in a straight line, to see people in far-off countries such as China or India?

A. Westward
B. Eastward
C. Upward
D. Downward

A. Westward? **B.** Eastward? **C.** Upward? **D.** Downward?

(Over)

Getting Ready

1. Photocopy Questionnaire. Make one copy of the *Pre–Unit 2 Questionnaire* for each student, plus one additional copy for each small discussion team of three or four students.

2. Obtain one Earth globe for each team of four students. These can be plastic inflatable globes or you can borrow Earth globes or use large balls to represent the Earth. If possible, remove the globes from their stands, and place them on bowls or on rolls of masking tape, so they will not roll off the tables.

Taking the *Pre–Unit 2 Questionnaire*

1. Introduce the unit. Tell the class that they will be studying space science, and that they will start by studying our own planet, the Earth.

2. Introduce the *Pre–Unit 2 Questionnaire.* Explain that each student will now receive a questionnaire about the Earth. Emphasize that usually it is fine for them to help one another, but this time you want them each to answer the questions without talking to anyone else. The questionnaire is designed to find out what each student thinks.

3. Make best guesses. Explain that they will probably not know all the answers. That's fine. Tell them that many adults would not know the answers to some of these questions. Mention that the word *gravity* is in the questionnaire, and if they don't know much about gravity, that's okay, as they will learn more soon. Advise the students to think about each question, and if they don't know the answer, make a good guess. Later, when they have learned more, they will get to fill out the questionnaire again to find out how their ideas have changed. Say that the questionnaire will not be used for a grade.

Unit Goals

Earth, like all planets and stars, is round like a sphere.

Earth's gravity pulls things toward the center of Earth.

Gravity is everywhere.

TEACHER CONSIDERATIONS

PRE–UNIT 2 QUESTIONNAIRE, PAGE 2

Session 2.1 Student Sheet Name ____________________

Pre–Unit 2 Questionnaire, Page 2

4. If a person is standing on the Moon holding a rock and then lets go of it, what will happen to the rock? It will __________

A. Float away.
B. Float where it is.
C. Fall to the ground.
D. Go up.

5. This drawing shows some enlarged people dropping rocks at various places around the Earth. What happens to each rock? Draw a line showing the complete path of the rock from the person's hand to where it finally stops. Why will the rock fall that way?

____

____

____

____

____

____

____

____

____

QUESTIONNAIRE CONNECTION

We have included *Questionnaire Connection* notes in later sessions of the unit to help you highlight when an activity or discussion relates to specific questions on the questionnaire. It is important to look over your students' questionnaires before getting too far into the unit, so that you can identify areas that need more or less instruction, and any major misconceptions students have.

Scoring Guide for the *Pre–Unit 2 Questionnaire:* The questionnaire can be scored using the scoring guide on page 74.

Different Order of Questions on the Pre and Post Questionnaires: The *Post–Unit 2 Questionnaire* in Session 2.6 is identical in content with the *Pre–Unit 2 Questionnaire,* but the placement of some questions and the order of the possible answers is different. This is to encourage students to answer thoughtfully, rather than simply to remember the order of the answers.

4. Explain what to do when they finish. You may want to tell students who finish early to read quietly so that everyone has a chance to do their best.

5. Distribute questionnaires. Give each student a copy of the *Pre–Unit 2 Questionnaire*. Mention that on question #4, the people in the drawing are supposed to be ordinary-sized people, even though they are drawn very large compared with the Earth. (Some students assume that these are giant people dropping rocks from space; this is not the intent of the question.)

6. Collect the questionnaires. When everyone is finished, collect the questionnaires, making sure that students have put their names on them. Explain that they will be discussing and learning about these questions in the days to come.

Conducting Small–Group Discussions

1. Designate small discussion teams. Organize the class into discussion teams of three to five students per team. Explain that each team will discuss each question and come to agreement, if possible, on the best answers. Say that everyone on a team should have a chance to say what they think the answers are and explain their thinking. Each team will receive one blank questionnaire to use for recording their team's final answers.

2. Explain and distribute blank questionnaires and globes. Say that each team will get an Earth globe, which might be helpful in pointing things out to one another as they discuss the questions. Tell students that the globe is really a model of the Earth, and remind them that scientists often use models to explain or better understand how things work. Tell them that they will get a roll of masking tape to set on their table to act as a base for the globe when they aren't using it. Let them know that they are not to toss or bounce the globes. Pass out one blank questionnaire to each team.

3. Circulate during discussion. Circulate among the teams of students, encouraging them to discuss any disagreements fully and to use the globes to demonstrate their ideas. Teams who agree on the answers early should make a list of arguments in support of their answers.

Unit Goals

Earth, like all planets and stars, is round like a sphere.

Earth's gravity pulls things toward the center of Earth.

Gravity is everywhere.

TEACHER CONSIDERATIONS

TEACHING NOTES

Refrain from providing correct responses now: Why not just tell students the correct answers? Research has found that students are often quite reluctant to let go of their old ideas unless they are confronted with evidence that refutes these ideas, and have time to process the evidence and new ideas. Simply giving the correct answers to questions such as these generally is not an effective strategy for changing students' ideas. Correct responses to the questions will be provided during the activities and discussion in the coming sessions. The discussion and investigation processes also help promote an atmosphere of inquiry in the classroom.

"Student understanding is actively constructed through individual and social processes, not simply by being told answers."—National Science Education Standards (NRC, 1996)

What Some Teachers Said

" This lesson was effective for introducing models of Earth. These students were able to grasp the concept upon the initial introduction of the models. The pre questionnaire review revealed some new understanding."

"It was very enlightening to see that there are many misconceptions about why the earth seems flat to us, and about what gravity is and how it works."

Conducting a Class Discussion

1. Lead the class in a discussion about the questionnaire. Play the role of moderator, requiring each group to support their ideas with arguments or to demonstrate, using the Earth globes.

2. Poll students after the discussion of each question. After discussing one question, poll the class on the alternative answers. Following are some suggestions for facilitating the discussion:

Question 1. For the teacher: The correct answer is "D. The Earth |is round like a ball, but it looks flat because we see only a small part of it."

Variation in student responses. You can expect some variation in your students' ideas on this question, because it requires a correct understanding of the Earth's scale and the part-to-whole relationship between the "flat ground" of our everyday experience, and the "ball-shaped Earth" that we learn about in school. For example, a student may think that the Earth we live on is really flat, and the ball-shaped Earth is a "a planet in the sky, where only astronauts go."

Question 2. For the teacher: The correct answers are "B. Gravity is an invisible force; D. There is a pull of gravity between all objects; F. Gravity keeps the Moon in Earth's orbit."

Variation in student responses. So far, the word *gravity* hasn't been defined. Tell students that they will be learning more about gravity in later sessions, but for now, don't define it yourself. During this discussion, listen to what students say about gravity, and informally assess whether they harbor common misconceptions, such as "Gravity is caused by air" or "There is no gravity in space."

Question 3. For the teacher: The correct answer is "D. Downward."

Looking in a straight line. When first confronted with this question, most people try to imagine which direction they would fly in a plane to get to Australia, and will answer, "Southward" or "Westward." Ask your students to imagine that the Earth is made out of glass and that they can look straight through it. You might also use a globe and a ruler to show what happens if you look due east or west: the ruler (representing the way you would look) points off into space. Ask a few supplementary questions, such as, "If you were in Antarctica, which way would you point toward North America?"

Unit Goals

Earth, like all planets and stars, is round like a sphere.

Earth's gravity pulls things toward the center of Earth.

Gravity is everywhere.

TEACHER CONSIDERATIONS

TEACHING NOTES

Classroom Management—when to collect the globes: During the whole-class discussion, it is convenient to have the Earth globes on student desks so they can illustrate points with them. However, if students cannot resist the temptation to play with the globes, collect them before the discussion, and keep one handy to demonstrate points.

Leave one globe inflated to use in the next session.

Adjusting for student experience: If your students have experienced Unit 1 of this sequence, or are otherwise familiar with the size of the Earth, they may be able to apply this knowledge to the discussion of Question 1. If they haven't yet learned about the scale of the Earth, there are some optional activities in Session 2.2 that may help.

See Page 283 for some tips on leading good discussions.

Key Vocabulary

Science and Inquiry Vocabulary

Evidence
Scientific Explanation
Model
Scale Model
Prediction
Scientist
Three–Dimensional (3-D)
Two–Dimensional (2-D)

Space Science Vocabulary

Atmosphere
Air Resistance
Force
Mass
Gravity
Weightless
Satellite
Orbit
Diameter
Sphere
System

Question 4. For the teacher: Rocks fall toward the surface. The correct answer shows each rock falling straight down, landing next to the person's feet. It is common for students to show the rocks falling off the Earth to an absolute down direction in space, or to mix the correct and the incorrect and depict the rock falling at an angle.

Draw circles and invite students to draw alternative answers. To help the students discuss their answers to this question, draw three or four large circles on the board, each with figures holding rocks, as shown on the questionnaire. Invite students to come up to the board to draw their answers. Ask them to explain why the rock would fall the way they have drawn it. The pictures of three or four alternative views will help you focus the discussion on which answer is best.

"Down" is always toward the center. At some point in the discussion, you may need to explain why "down" is always toward the center of the Earth. Ask your students to think about the people who live all around the ball-shaped Earth. The only way to explain why these people do not fall off is to imagine that "down" is toward the center of the Earth. To demonstrate this idea, turn an Earth globe so that the South Pole is "up," and ask the students to imagine being there. People on the South Pole must think that people in the Northern Hemisphere live upside-down!

Question 5. For the teacher: The correct answer is "C. Fall to the ground."

Unit Goals

Earth, like all planets and stars, is round like a sphere.

Earth's gravity pulls things toward the center of Earth.

Gravity is everywhere.

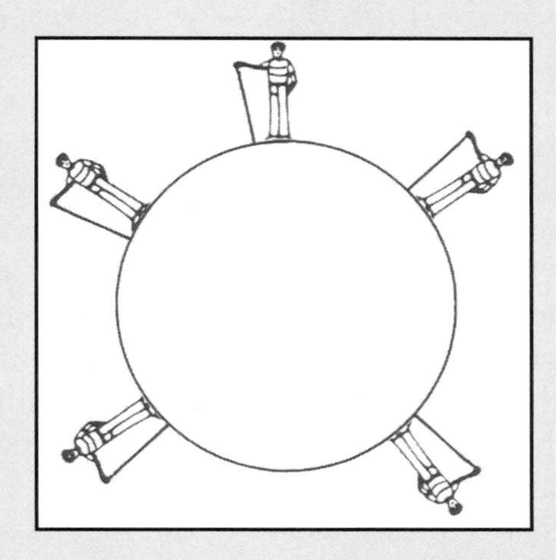

OPTIONAL PROMPT FOR WRITING OR DISCUSSION

You may want to have students use one of the prompts below for science journal writing at the end of this session or as homework. The prompts could also be used for a discussion or during a final student sharing circle.

- Right now, I'm not sure about. . .
- Some evidence I've never thought about before is. . .
- A good example of evidence that I heard today is. . .
- I changed my mind about this today, because. . .

Key Vocabulary

Science and Inquiry Vocabulary

Evidence
Scientific Explanation
Model
Scale Model
Prediction
Scientist
Three–Dimensional (3-D)
Two–Dimensional (2-D)

Space Science Vocabulary

Atmosphere
Air Resistance
Force
Mass
Gravity
Weightless
Satellite
Orbit
Diameter
Sphere
System

SESSION 2.2
What Shape Is the Earth?

Overview

The session begins with an introduction to key concepts about models. In a large–group discussion, students examine two different models of the Earth: a spherical and a flat-Earth model. They discuss why the flat-Earth model might have made sense to people in ancient times.

Students learn that the spherical Earth model is based on *evidence*. After defining how scientists use evidence, the class embarks on a virtual orbit around the Earth, through a series of satellite images. They observe that the Earth looks flat from close up, but the farther you get away from it, the easier it is to see its spherical shape. After their orbital tour, the class reviews two questions about the Earth's shape from the *Pre–Unit 2 Questionnaire*. Three key concepts about Earth's shape are posted on the class concept wall.

The students then work in small groups called *evidence circles* to respond to four statements by (fictional) people who believe that the Earth is flat. Students get practice in using evidence-based arguments, and in the process, they may solidify their own and their fellow students' conceptions about the Earth.

What Shape Is the Earth?	Estimated Time
Introducing models	5 minutes
Class discussion of two models of earth's shape	15 minutes
Taking a *Trip Around the Earth*	10 minutes
Review questionnaire and post space science key concepts	10 minutes
Evidence Circles: *What Would You Say to Them?*	20 minutes
TOTAL	**60 minutes**

What You Need

Note: *In this session, the class discusses two models of the Earth's shape, a globe and a "flat–Earth" model. You can illustrate the two models simply by using the* Comparing Earth Models *transparency from the transparency packet. However, we recommend that you compare a globe with a three–dimensional model similar to one advocated by the Flat Earth Society. See Getting Ready #1, which follows for details.*

Unit Goals

Earth, like all planets and stars, is round like a sphere.

Earth's gravity pulls things toward the center of Earth.

Gravity is everywhere.

TEACHER CONSIDERATIONS

ASSESSMENT OPPORTUNITY

Critical Juncture–Understanding the Earth's Spherical Shape: Before moving on to explore Earth's gravity in Session 2.3, it is important that your students have an understanding of the Earth's spherical shape. By listening to their discussions in this session, you may note whether any further experience is needed. The first two ideas in *Providing More Experience* below may help students who have not yet grasped the scale of the Earth and how that relates to our perceptions of its shape. Discussing the Paper Plates model in #3 may help students refine and solidify their mental models of the spherical Earth and how the evidence supports that model.

PROVIDING MORE EXPERIENCE

1. **Insect on Globe:** Hold up the globe and tell your students to imagine an insect even smaller than an ant crawling on it. Tell them that the globe would probably look flat to the insect. No matter where it crawled on the globe it would still look flat. Tell them that's because the globe is so big compared to the insect. The insect can only see a small part of it at a time. It's when you see the whole thing that you see the spherical shape. Tell them that people are much smaller compared to the Earth than an insect is to the globe. The Earth is so big compared to people that it looks flat to us, even from up high in an airplane. To see its roundness, you have to move even farther away from it.

2. **Nose Against a Building:** Tell your students to imagine they were led blindfolded to a very large building. Tell them to imagine that they were then placed with their nose against an outside wall of the building and the blindfold removed. What if they were then asked to describe the shape of the building from this position? To see the shape of the building, they'd have to step back away from it. To see the spherical shape of the Earth, which is much larger, you have to move much farther away from it.

3. **Paper Plates:** Students could look at and consider paper plates as a model of a "flat but round" Earth. Hold up a plate and move your finger in an orbit around it, from front to back. Ask them to imagine that your finger is a spacecraft orbiting the plate. Point out that from the spacecraft the Earth would look round when seen from the front or back, but it would look like a line when viewed edgewise, from two vantage points every orbit. Contrast this with your finger orbiting a globe, and point out that the Earth would appear round from every point of the orbit. Tell them that the Earth is not shaped round like a paper plate. The evidence is that from space it looks round from every perspective, like a ball.

Key Vocabulary

Science and Inquiry Vocabulary

Evidence
Scientific Explanation
Model
Scale Model
Prediction
Scientist
Three–Dimensional (3-D)
Two–Dimensional (2-D)

Space Science Vocabulary

Atmosphere
Air Resistance
Force
Mass
Gravity
Weightless
Satellite
Orbit
Diameter
Sphere
System

For the class

- ❑ overhead projector or computer with large screen monitor/LCD projector
- ❑ overhead transparency from the transparency packet, or CD–ROM file of *Comparing Earth Models* or 2 3–D models of the Earth (see *Getting Ready,* which follows), for which you will need:
  1 Earth globe (ideally opaque, rather than transparent)
  1 or 2 identical large clear plastic or glass bowls
  1 copy of round *Earth Map with Stars, Moon, Planets, and Sun* from the student sheet packet
- ❑ overhead transparency or CD–ROM file of questions #1 and #3 from the *Pre–Unit 2 Questionnaire*, from the transparency packet
- ❑ CD–ROM file or overhead transparencies of 15 *Orbit Around the Earth* photos from the transparency packet
- ❑ 1 ruler
- ❑ 1 pair of scissors
- ❑ clear adhesive tape
- ❑ 12 sentence strips for key concepts
- ❑ wide-tip felt pen

For each group of four students

- ❑ 1 piece of scratch paper
- ❑ 1 copy of the two–page *What Would You Say to Them?* sheet from the student sheet packet
- ❑ pencils

For each student

- ❑ *optional: Laika: the First Astronaut* reading (on CD–ROM)

Getting Ready

1. Decide which models to use. In this session, the class will discuss two different models of the Earth's shape. You can conduct the discussion by using the *Comparing Earth Models* transparency, which has drawings of the two models, or you can use 3–D models. For the 3–D version, an Earth globe from Session 2.1 can serve as Model A. You'll need to construct a 3–D Model B, the flat–Earth model. (See directions for making a flat–Earth model on pages 251 and 253.)

2. Prepare to display the images for the *Trip Around the Earth*, using the CD–ROM or overhead transparencies.

Unit Goals

Earth, like all planets and stars, is round like a sphere.

Earth's gravity pulls things toward the center of Earth.

Gravity is everywhere.

TEACHER CONSIDERATIONS

TEACHING NOTES

Why use the flat-Earth model at all? Model B is based on a model advocated by the Flat Earth Society, a group that does not accept that the Earth is spherical, despite all the evidence supporting a spherical Earth. Model B is also similar to some models used in ancient times and to models sometimes described by young children. It is included here to give students an opportunity to notice and verbalize what is wrong with it. Any students in your class who might privately harbor aspects of this model in their own mental models will have a chance to recognize its many limitations and inaccuracies, and replace it with a more accurate model.

To make a 3-D Model B:

a. Make a copy of the two-piece round, flat-Earth map from the student sheet packet and cut out the small drawings of stars, Moon, planet, and Sun in the margins of the map. Glue or tape them to the inside surface of a transparent bowl.

b. Cut out the two halves of the map, tape them together, and trim the map to fit the circumference of your bowls. (It's okay if some continents are trimmed in the process.)

c. Place the round, flat-Earth map face up, so it's covering the top of one clear bowl.

d. Place the second clear bowl upside-down over the map. *Note: If you have only one bowl, you can eliminate the bowl on the bottom.*

An optional student reading: An optional reading is provided on the CD-ROM that relates to the concepts presented in this and other sessions of the unit. The reading, *Laika: The First Astronaut*, is about the first living Earth creature sent into space: a dog. The reading introduces ideas about gravity and weightlessness, which will be taught later in the unit. More information about the goals of the reading and suggested discussion questions are included on the CD-ROM, along with the reading.

***Special note*—The Laika reading may be disturbing:** Please review the reading before assigning it to your class, to determine if it would be disturbing for your students. There were no plans in the mission to return Laika to Earth, and she died about seven hours into the mission.

LAIKA—THE FIRST ASTRONAUT

Session 2.2 Optional Student Reading

Laika: The First Astronaut

People explored the skies of Earth for many years. First, they explored in balloons, and then, later, in planes. They found that the higher you went, the less air there is. Where there was less air, they found that they needed air tanks to live. But no one had traveled higher, where there is no air at all. No one had traveled above the atmosphere into space. Scientists were afraid that people's bodies might have trouble in space, and they might not be able to live. They were afraid to send the first astronaut into space.

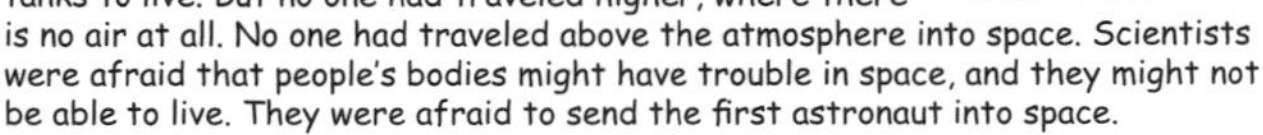

The first astronaut was found living on the streets. She had a long tongue that hung out of her mouth. Her body was covered with fur. When she was happy, she wagged her tail. The first to travel into space was Laika, a dog.

Laika went through lots of training with other dogs. She learned to stay still in a spaceship for many days. She learned how to eat special foods.

Scientists wanted evidence that it is possible to live in space before sending up a person.That's why they sent up dogs as the first astronauts. Other dogs had been in rocket ship tests, but none had been into space before Laika.

Laika was put in a tiny spaceship with air and food. It was four meters high.The spaceship was on a big rocket to help carry it high into space, where there is no air.

The rocket carried Laika 1,500 kilometers high above the Earth. That is so high that it is above the atmosphere. It started orbiting the round Earth. Above the air, her spaceship could travel very fast. It traveled so fast, it could orbit the Earth in 1 hour and 45 minutes!

Laika never returned to Earth. She died about seven hours after take-off. After she died, her spaceship kept orbiting Earth for 162 days. It orbited the Earth 2,570 times. After that, her spaceship moved back down to where there is air. Her spaceship was traveling so fast when it went into the atmosphere that it got very hot. It got so hot that it burned up.

Many people were very sad when Laika died in space. Many other dogs and monkeys made trips into space, and returned to Earth healthy. These animals helped make space flight safe for the people who followed them.

3. If you will not be using the CD-ROM, make an overhead transparency of each of the following pages:
- Questions #1 and #3 from the *Pre–Unit 2 Questionnaire*
- 15 *Trip Around the Earth* images

4. For each team of four students, photocopy the two-page *What Would You Say to Them?* sheet. Cut each sheet in half, and organize them so that each team will have all four statements.

5. *Optional*: Make one copy per student of the reading *Laika: The First Astronaut,* which can be found on the CD-ROM.

6. Choose a wall or bulletin board to serve as a "concept wall" for Unit 2. This is where you will post sentence strips showing the key concepts. You'll post the strips in two columns. (Please see the illustration on page 327, at the end of this unit.) The left column should have a title that says, *What We Have Learned About What Scientists Do*. The right column should be titled, *What We Have Learned About Space Science.*

7. Prepare sentence strips for the following key concepts introduced during this session. Have them ready to post during the session under the *What We Have Learned About Space Science* side of the concept wall:

The Earth, like all planets and stars, is round, like a sphere.
The Earth is so big that we can't see that it's round when we're on it.
People live all over the surface of the Earth.

8. Before the session, post the following nine key concepts on the *What We Have Learned About What Scientists Do* side of the concept wall. Your introduction of these concepts in Session 2.2 will be brief, but these concepts are woven throughout Unit 2 and the entire Space Science Sequence, and students will revisit them.

Unit Goals

Earth, like all planets and stars, is round like a sphere.

Earth's gravity pulls things toward the center of Earth.

Gravity is everywhere.

TEACHER CONSIDERATIONS

TEACHING NOTES

On the CD-ROM there are 29 images for the *Trip Around the Earth* activity. If you use transparencies to show the images instead of the CD-ROM, you will have 15 images, which are adequate for the purpose of taking off and making a full orbit.

Putting together Model B

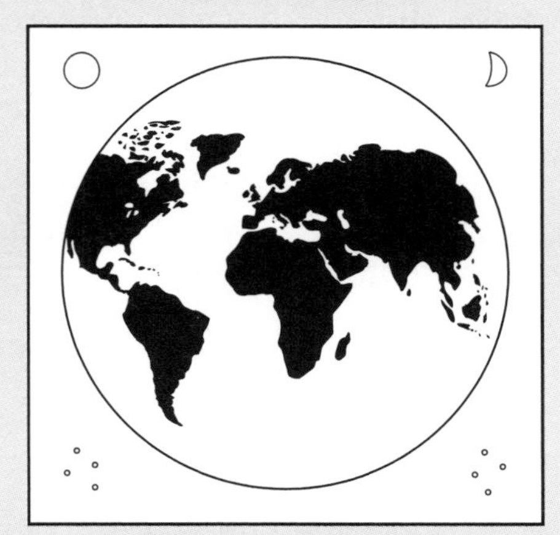

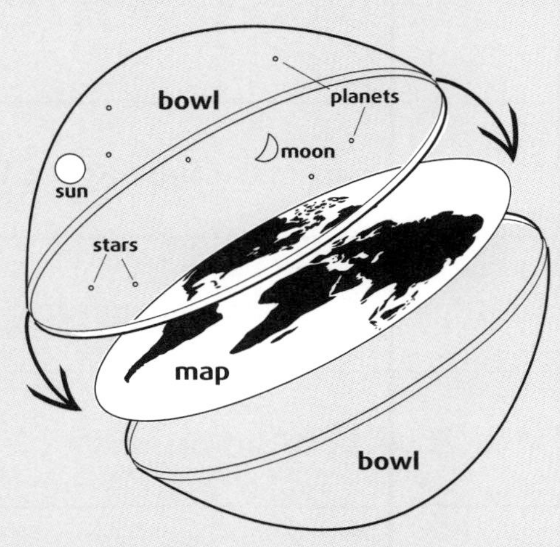

CD-ROM NOTES

CD–ROM version of the Trip Around the Earth: If you are using the CD–ROM version, preview it before class. Explore the curvature of the Earth by taking the multimedia trip up through the atmosphere and then around the Earth. Starting at the Kennedy Space Center in Florida, where the space shuttle is launched, you can quickly rise though the atmosphere by clicking the "up" arrow or the altitude buttons with your mouse. Each picture correlates with the altitude button highlighted on the right. Click the "down" button with your mouse to go back toward the Earth. To enter "orbit," click the "around" button next to the "up" and "down" buttons. Once you are in "orbit," you can use the "frame" arrow to circumnavigate the Earth, frame by frame, from space. Also, try clicking the "Spin" button to see the Earth rotate. Further instructions for using this program are included on the CD–ROM.

What Some Teachers Said

"The CD really wowed my students. Even though it was probably a model of the earth, the students were convinced of the curvature of the earth's surface."

"The students were very intrigued with the CD and the views of Earth from different distances."

"The successive transparencies were a great way to help students grasp how the curvature of the planet becomes apparent as you gradually get further away from it. I think it really helped them understand that even though they see the Earth as flat, it is really curved."

"The CD-ROM played a big part in my students grasping the idea of why the Earth appears to be flat in one picture, but round in another. The discussion about evidence was fun. Students were very firm in their explanations."

If you have presented Unit 1, note that all nine of these concepts about evidence and models were presented during that unit, and should remain on the concept wall. If this is your first unit in the sequence, write the key concepts below on sentence strips, and post them on the concept wall before the session.

Six of the concepts introduced in this session are about models:

Scientists use models to help understand and explain how things work.
Space scientists use models to study things that are very big or far away.
Models help us make and test predictions.
Every model is inaccurate in some way.
Models can be 3–dimensional or 2–dimensional.
A model can be an explanation in your mind.

Three of the concepts introduced in this session are about evidence:

Evidence is information, such as measurements or observations, that is used to help explain things.
Scientists base their explanations on evidence.
Scientists question, discuss, and check one another's evidence and explanations.

Unit Goals

Earth, like all planets and stars, is round like a sphere.

Earth's gravity pulls things toward the center of Earth.

Gravity is everywhere.

TEACHER CONSIDERATIONS

TEACHING NOTES

Many concepts: students who did not encounter them in Unit 1 are introduced to quite a few new key concepts about models and evidence in this session, along with the three concepts about the Earth's shape. Reassure students that they will have further opportunities to understand these concepts: they don't need to absorb them all at once.

Key Vocabulary

Science and Inquiry Vocabulary

Evidence
Scientific Explanation
Model
Scale Model
Prediction
Scientist
Three–Dimensional (3-D)
Two–Dimensional (2-D)

Space Science Vocabulary

Atmosphere
Air Resistance
Force
Mass
Gravity
Weightless
Satellite
Orbit
Diameter
Sphere
System

Introducing the Concept Wall

1. Scientists often use models. Say that scientists use *models* to explain things they have observed, show how they think things in the natural world work, make predictions, or learn more about things they can't look at directly.

2. Six key concepts about models. Read the six concepts on the concept wall about models with the students, and clarify any terms they may not have encountered earlier. Tell them that they will be using models and referring to these key concepts during the coming activities.

Scientists use models to help understand and explain how things work.
Space scientists use models to study things that are very big or far away.
Models help us make and test predictions.
Every model is inaccurate in some way.
Models can be 3–dimensional or 2–dimensional.
A model can be an explanation in your mind.

Large–Group Discussions of Two Models of Earth's Shape

1. Introduce two models of Earth's shape. Tell the class that they will be discussing two models of the Earth. Introduce the globe as Model A and the two-bowl model you prepared earlier as Model B. (Or show the two models on the *Comparing Earth Models* transparency or CD-ROM file.) Students are already familiar with Model A. Describe Model B, saying that the Earth's shape is flat and round in this model, with a dome over it. From the outside, it looks ball-shaped. The Sun, stars, and Moon are on the inside surface of the dome.

Unit Goals

Earth, like all planets and stars, is round like a sphere.

Earth's gravity pulls things toward the center of Earth.

Gravity is everywhere.

TEACHER CONSIDERATIONS

TEACHING NOTES

Adjusting for student experience: If this is your students' first encounter with models, you might take some extra time to explain how scientists use models, as is done in Unit 1, Session 1.4 of the *Space Science Sequence*.

1. **Introduce models.** Hold up a scale model car (or any model), and say that a model is something that shows or explains what the real thing is like.

2. **All models are different from the real thing in at least one way.** Emphasize that good models are like the real thing, but no model is exactly the same as the real thing. Ask, "What are some ways the model car is not exactly the same as a real car?" [It's smaller, it has no motor, the doors don't open, the tires are metal, it doesn't have gas in it, it has no lights, it can't move under its own power, and so on.]

Note: Students enjoy finding inaccuracies in models, and they are usually good at it!

3. **Define scale models.** Tell them that although it's much smaller than a real car, this model looks like a car because someone measured every part and made them all smaller by the same amount. It is a *scale model* of a real car.

4. **Compare scale models.** Hold up a model truck or other vehicle in the same scale as the car. Say, "Someone shrank the measurements for this truck down exactly the same amount as the car. We can see how the sizes of the car and truck compare with each other by looking at these scale models."

5. **A model can also be an idea in your mind.** Mention that sometimes you have a picture in your mind about how something works. This is a kind of model, too: someone's mental explanation of how something looks or works.

6. **Scientists often use models.** Say that scientists use all kinds of models, to explain things they have observed, show how they think things in the natural world work, make predictions, or learn more about things they can't look at directly. Space scientists often use scale models because many of the things they study are very large and far away.

Key Vocabulary

Science and Inquiry Vocabulary

Evidence
Scientific Explanation
Model
Scale Model
Prediction
Scientist
Three Dimensional (3-D)
Two Dimensional (2-D)

Space Science Vocabulary

Atmosphere
Air Resistance
Force
Mass
Gravity
Weightless
Satellite
Orbit
Diameter
Sphere
System

2. All models are inaccurate in some ways. Say that these two models represent two different ideas about how the Earth is shaped. Say that neither of the models is exactly like the real Earth. Ask students to brainstorm ways in which *both* models are inaccurate.

They may say:
- They are much smaller than the real Earth.
- They are made of different materials than the Earth.
- They are not moving like the Earth. (and so on)

3. Model A is accurate because it is *spherical.* Focus their attention on Model A. Ask them what is accurate about the model. [It shows the Earth as round like a ball.] Introduce the word *sphere* (and *spherical*, meaning "ball–shaped"), and say that this is the model that scientists accept.

4. Why might ancient people think that Model A was not a good model? Tell them to imagine that they were people who lived on Earth a couple of thousand years ago, and did not know that the Earth is shaped like a sphere. They might think that Model A was a ridiculous model, because the Earth looks flat when you are on it. Say that many people back then did think that the Earth was flat.

5. Why might ancient people think that Model B was a good model? Now, focus their attention on Model B. Tell them to imagine why people who didn't know the Earth is spherical might think that this was a good model.

Students may say:
- From the perspective of people on Earth, the Earth looks flat.
- Looking up from Earth, the sky looks like a dome.
- From Earth, the Moon, stars, and Sun look as though they are all about the same distance away from Earth on the inside of the dome.
- From Earth, the Moon and Sun look as though they are the same size.
- From Earth, the stars and planets look as though they are the same size.

6. Model B shows a round Earth that looks flat from the surface. Tell students that some people who have heard that the Earth is a sphere might think that this is a good model, because, for them, it explains how Earth can look like a sphere from the outside, but look flat when you're on it.

Unit Goals

Earth, like all planets and stars, is round like a sphere.

Earth's gravity pulls things toward the center of Earth.

Gravity is everywhere.

COMPARING EARTH MODELS

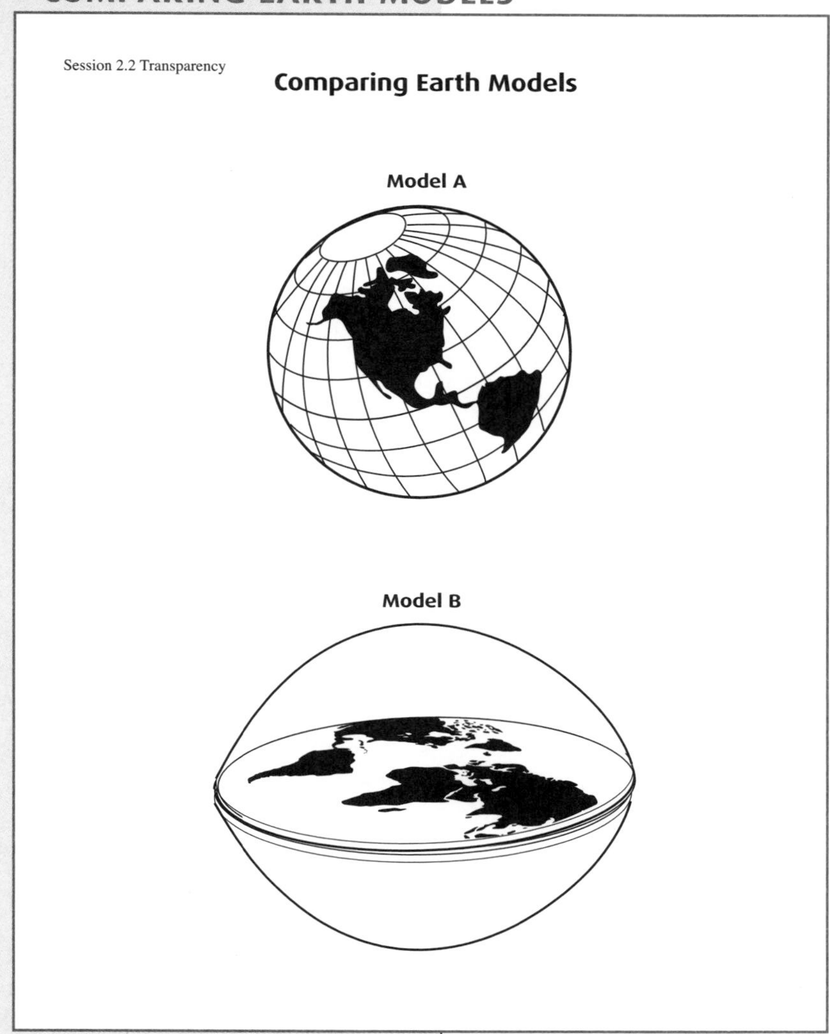

What One Teacher Said

"Students engaged in lively discussions, both in small groups and with the whole class. They asked pertinent questions and shared interesting reasoning."

7. Introduce *evidence*. Explain that scientists look for clues to help explain things. These clues are called *evidence*. Read the three key concepts about evidence on the concept wall.

Evidence is information, such as measurements or observations, that is used to help explain things.
Scientists base their explanations on evidence.
Scientists question, discuss, and check one another's evidence and explanations.

8. Why is Model B not a good model? Now, tell them to evaluate Model B from the perspective of modern people who have evidence about the shape of the Earth. Ask them to list some evidence that Model B is not a good model.

They may say:

- The real Sun, Moon, planets, and stars are not the same distance away from Earth.
- The real Sun, stars, and some planets are much bigger than the Earth. They could not fit on the dome "inside the Earth."
- The real Earth is round like a solid ball, not a ball that is mostly air.
- We live on the surface of the Earth, not inside it.

9. Evidence shows that Model B is not accurate. Set Model B to the side, and tell them that they have listed evidence that shows that it is not a good model. Hold up Model A, and tell them that the spherical Earth model is accepted by scientists because it does match all the evidence.

10. Evidence that the Earth is round like a sphere. If no one has mentioned it, tell them that one piece of evidence that the Earth is spherical is that people have seen Earth's shape from space. Tell them that they will now be taking a virtual trip around the Earth by looking at pictures.

Unit Goals

Earth, like all planets and stars, is round like a sphere.

Earth's gravity pulls things toward the center of Earth.

Gravity is everywhere.

TEACHER CONSIDERATIONS

Key Vocabulary

Science and Inquiry Vocabulary

Evidence

Scientific Explanation

Model

Scale Model

Prediction

Scientist

Three Dimensional (3-D)

Two Dimensional (2-D)

Space Science Vocabulary

Atmosphere

Air Resistance

Force

Mass

Gravity

Weightless

Satellite

Orbit

Diameter

Sphere

System

Taking a Trip Around the Earth

1. Orbit the Earth. Remind students that making a trip around the Earth in space is called *orbiting* the Earth. Ask if they can think of other things that orbit the Earth. [Satellites, the Moon.]

2. Begin showing the images. Show the first few images of the *Trip Around the Earth* on CD-ROM or transparencies. As you "take off" from Earth and gain altitude, ask, "What shape does the Earth look like in this picture?" [Flat.] When you get to the image in which students can see a very slight curvature of the Earth's horizon [image 6 in the transparencies, 7 on the CD-ROM], ask the same question. Ask students to raise their hands if they think that it looks a) curved, or b) flat.

3. Show curvature. If the curvature is still hard to see, set a ruler or meter stick on the edge of the Earth in the photo. Point out that compared with the ruler, the edge of the Earth is slightly curved.

4. Easier to see curve with distance. In the next photo, point out that the curve is more obvious. Ask why this might be. [The farther you get away from Earth, the easier it is to see that it actually is round like a sphere.]

5. Evidence the Earth is round. Show the last five images to simulate completing one orbit of the Earth. Point out that as the class orbits the Earth, they have seen evidence that it is round from every side and angle, like Model A. Point out that it does not look anything like Model B.

6. Watch the ball-shaped Earth spin. If you are using the CD-ROM, make the Earth spin. If you are using overhead transparencies, spin a globe. Point out that the Earth appears round from every direction.

7. The Moon is also spherical. Ask what shape the Moon is. [Round.] Ask, "Is it round like a plate or round like a ball?" Tell them that astronauts have also orbited the Moon, and have evidence that it is a sphere.

8. The Sun is also a sphere. Tell them that we have evidence that the Sun is shaped like a sphere, because planet Earth orbits the Sun. From everywhere in our orbit, the Sun looks round.

Unit Goals

Earth, like all planets and stars, is round like a sphere.

Earth's gravity pulls things toward the center of Earth.

Gravity is everywhere.

TRIP AROUND THE EARTH

What One Teacher Said

"My students were totally engaged in the discussions and especially liked the overhead transparencies of the earth from space. I think they will have a new appreciation for maps and globes now that they have seen the world from space. They wanted to discuss more about space flight and how the rocket works."

Key Vocabulary

Science and Inquiry Vocabulary

Evidence

Scientific Explanation

Model

Scale Model

Prediction

Scientist

Three Dimensional (3-D)

Two Dimensional (2-D)

Space Science Vocabulary

Atmosphere

Air Resistance

Force

Mass

Gravity

Weightless

Satellite

Orbit

Diameter

Sphere

System

Review the Evidence for Earth's Spherical Shape

1. Lead a discussion about Questionnaire question #1. Show the transparency or CD-ROM file of questions #1 and #3. On the first question, go through options A–E, and ask them to explain what's wrong (or right) with each choice. [They should arrive at the conclusion that D is the best answer.]

2. Discuss question #3. Discuss the alternatives. If students are still confused about whether "Eastward" or "Westward" might be the correct answer, place a ruler on the globe, as before, to demonstrate that moving eastward or westward in a straight line would take you off the globe and out into space. Ask a student to use the globe to demonstrate why "D" is the correct answer.

Post the Key Concepts About Earth's Shape

1. Tell students that they have learned some space science concepts that are important to remember. Post the three concepts under *What We Have Learned About Space Science.*

The Earth, like all planets and stars, is round, like a sphere.
The Earth is so big that we can't see that it's round when we're on it.
People live all over the surface of the Earth.

Small-Group Discussions: *What Would You Say to Them?*

1. Scientists discuss evidence from investigations. Say that they just gathered evidence about the Earth's shape. Let students know that one way scientists do their work is to discuss the evidence from their investigations. They listen to one another, ask questions, present evidence, argue, and try to agree about what is true—what explanation best matches *all* available evidence.

Unit Goals

Earth, like all planets and stars, is round like a sphere.

Earth's gravity pulls things toward the center of Earth.

Gravity is everywhere.

QUESTIONNAIRE QUESTIONS 1 AND 3

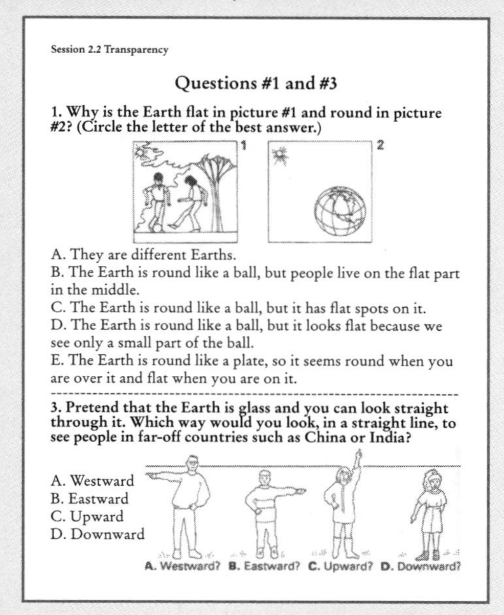
Session 2.2 Transparency

Questions #1 and #3

1. Why is the Earth flat in picture #1 and round in picture #2? (Circle the letter of the best answer.)

A. They are different Earths.
B. The Earth is round like a ball, but people live on the flat part in the middle.
C. The Earth is round like a ball, but it has flat spots on it.
D. The Earth is round like a ball, but it looks flat because we see only a small part of the ball.
E. The Earth is round like a plate, so it seems round when you are over it and flat when you are on it.

---

3. Pretend that the Earth is glass and you can look straight through it. Which way would you look, in a straight line, to see people in far-off countries such as China or India?

A. Westward
B. Eastward
C. Upward
D. Downward

WHAT WOULD YOU SAY TO THEM?

Session 2.2 Student Sheet

Name__________

1. What would you say to them?

"The Earth is flat. I live in Kansas, and every direction I turn, it looks flat."

Name__________

2. What would you say to them?

"I live in Australia. The Earth is flat. If the Earth were round like a ball, then we would be hanging upside down by our feet in Australia. We live on top of a flat Earth, just like the people in the U.S. do."

Session 2.2 Student Sheet

Name__________

3. What would you say to them?

"The Earth is round like a plate. It is round, but flat."

Name__________

4. What would you say to them?

"There are two Earths. The Earth I live on is flat. The Earth astronauts go to is round."

Key Vocabulary

Science and Inquiry Vocabulary

Evidence
Scientific Explanation
Model
Scale Model
Prediction
Scientist
Three Dimensional (3-D)
Two Dimensional (2-D)

Space Science Vocabulary

Atmosphere
Air Resistance
Force
Mass
Gravity
Weightless
Satellite
Orbit
Diameter
Sphere
System

2. Small–groups of students. Say that they will be working again in their small–groups. The teams will be called *evidence circles,* because they will discuss statements based on evidence. Have each team number off, from one to four.

3. Four statements. Say that they will each get to read a statement by someone who doesn't understand that the Earth is shaped like a sphere.

4. Think of evidence and arguments. Read the first statement as an example: *"The Earth is flat. I live in Kansas, and every direction I turn, it looks flat."* On each team, student #1 will get a piece of paper with this statement on it, and read it aloud. Then each team will discuss all the evidence and arguments they can think of to convince the person from Kansas that the Earth is not really flat. Student #1 will write all the team's evidence and arguments on the paper.

5. Small teams discuss all four statements. Explain that students #2, #3, and #4 will each read their statement, the team will discuss it, and the student who holds the statement will write down the evidence and arguments.

The four statements:

1. The Earth is flat. I live in Kansas, and every direction I turn, it looks flat.

2. I live in Australia. The Earth is flat. If the Earth were round like a ball, then we would be hanging upside down by our feet in Australia. We live on top of a flat Earth, just like people in the U.S. do.

3. The Earth is round like a plate. It is round, but flat.

4. There are two Earths. The Earth I live on is flat. The Earth that astronauts go to is round.

6. Class discussion. When students have finished, ask one group to read aloud their responses to statement #1, and ask if other groups have anything to add. Do this for each statement. If students have trouble addressing any of the statements, spend a little time discussing it as a class.

Unit Goals

Earth, like all planets and stars, is round like a sphere.

Earth's gravity pulls things toward the center of Earth.

Gravity is everywhere.

ASSESSMENT OPPORTUNITY

Optional Embedded Assessment: After groups of students record their team's responses to the four questions, you may want to provide your students with the opportunity to individually write an argument using evidence to convince the fictional writer that the Earth really is spherical. Students' written responses to the first statement (*"The Earth is flat. I live in Kansas, and every direction I turn it looks flat"*) are an opportunity to evaluate individual student's abilities to write a coherent and compelling argument based on evidence from key concepts presented in Sessions 2.1 and 2.2. Completing this assessment would entail some extra class time or could be assigned as homework. Student work can also be assessed with the general rubrics provided on page 66.

Rubric for Embedded Assessment:

Students write arguments against statement 1 based on evidence.

	Understanding Science Concepts They key science concepts for this assessment are the following: 1. The Earth, like all planets and stars, is round like a sphere. 2. The Earth is so big that we can't see that it's round when we're on it. 3. People live all over the surface of the Earth.
4	*The student demonstrates a complete understanding of the science concepts and uses scientific evidence to support the written explanation.* Some possible pieces of evidence that could be used to support their arguments include the images of Earth from outer space that show the curvature of the Earth and the continuous nature of the Earth and the ability to go around the Earth from one point to another and end up at original destination.
3	*The student demonstrates a partial understanding of the science concepts.* Although understanding is demonstrated, the student does **not** tie all of these concepts together in the explanation and does not support the explanation with evidence from class activities. The "evidence" may be more generally experiential or rely on statements of presumed fact, rather than being drawn directly from classroom science experiences.
2	*The student demonstrates an insufficient understanding of the science concepts.* The student demonstrates an understanding of one or two of the key concepts but does not demonstrate an understanding of all of the concepts and does not use evidence from class to support the explanation.
1	*The content information is inaccurate.* Some possible inaccuracies are 1. In Kansas, you live on top of a flat Earth. 2. The Earth is flat on top where you live in Kansas and round elsewhere. 3. The Earth is round like a plate. 4. There are two Earths. Where you live in Kansas is flat, and the one that the astronauts see from space is round.
0	*The response is irrelevant or off topic.*
n/a	*The student has no opportunity to respond and has left the question blank*

SESSION 2.3
Gravity

Overview

The goal of this session and of subsequent sessions in Unit 2 is not for students to gain a complete understanding of gravity. Rather, the goal is to give students a general sense of gravity's effects on our planet and in space, to introduce related key concepts about gravity, and to address some common misconceptions about gravity.

The session begins with a quick demonstration and an introduction to some basic concepts about gravity. Students then look at photographic evidence of how gravity affects people on different parts of the Earth. If students have traveled to other parts of the globe, they contribute their own evidence about the pull of gravity in those locations. With this evidence in mind, students revisit question #4 from the Unit 2 questionnaire, which asks what direction rocks would fall on various parts of the Earth. The concept that the gravitational pull between the Earth and rocks pulls all the rocks toward the center of the Earth leads to a discussion of how we can "beat" gravity.

In small–group evidence circles, students discuss the effects of gravity by answering the question, "Is gravity strong or weak?" Evidence circles in this and other sessions in the *Space Science Sequence* help students practice scientific inquiry skills. Their discussions give students a chance to help one another understand challenging concepts, such as the nature of gravity. Once back in the large group, they share their evidence for whether gravity is weak or strong.

Students then read about the first person ever to orbit the Earth in space. They hear Yuri Gagarin's eyewitness description of the spherical Earth. They learn that although he felt weightless in his spaceship while in orbit, Gagarin did not escape the pull of gravity between himself and the Earth.

Gravity	Estimated Time
Introducing gravity	10 minutes
Revisiting questionnaire and *Beating Gravity*	10 minutes
Evidence circles and large–group discussion: *Is Gravity Strong or Weak?*	20 minutes
Reading and Discussing *Yuri Gagarin—First Person in Space*	20 minutes
TOTAL	**60 minutes**

Unit Goals

Earth, like all planets and stars, is round like a sphere.

Earth's gravity pulls things toward the center of Earth.

Gravity is everywhere.

TEACHER CONSIDERATIONS

TEACHING NOTES

Don't mention that this session is about gravity until your introduction, when it will be revealed with the pen and book "trick."

What One Teacher Said

"They were totally engaged in the discussion. They had a pretty good understanding of gravity before we discussed it, but had a much more clear understanding after the discussion."

Key Vocabulary

Science and Inquiry Vocabulary

Evidence

Scientific Explanation

Model

Scale Model

Prediction

Scientist

Three–Dimensional (3-D)

Two–Dimensional (2-D)

Space Science Vocabulary

Atmosphere

Air Resistance

Force

Mass

Gravity

Weightless

Satellite

Orbit

Diameter

Sphere

System

What You Need

For the class

- ❑ overhead projector or computer with large screen monitor/LCD projector
- ❑ overhead transparency or CD-ROM file of *Photos of People on Earth* from the transparency packet
- ❑ wide-tip felt pen
- ❑ 1 book
- ❑ 9 sentence strips for key concepts

For each student

- ❑ 1 copy of student reading *Yuri Gagarin–First Person in Space,* from the student sheet packet
- ❑ *optional:* 1 sheet of lined paper

Getting Ready

1. Read the information for the teacher about gravity and weightlessness beginning on page 34 in the *Background* section. Decide how much of this information will be appropriate to share with your class during the activity.

2. Have a pen and a book handy when you introduce the session.

3. If you will not be using the CD-ROM, make an overhead transparency of *Photos of People on Earth.*

4. Make a copy of the two-page reading *Yuri Gagarin—First Person in Space* for each student.

5. Arrange for the appropriate projector format (computer with large screen monitor, LCD projector, or overhead projector) to display images to the class.

Unit Goals

Earth, like all planets and stars, is round like a sphere.

Earth's gravity pulls things towards the center of Earth.

Gravity is everywhere.

TEACHER CONSIDERATIONS

TEACHING NOTES

Reading level: The reading level of page 1 of the reading is appropriate for most third and fourth graders. Page 2 and the additional pages on the CD-ROM are for students who are interested in further information on the topic, and who are able to read at a slightly higher level.

Key Vocabulary

Science and Inquiry Vocabulary

Evidence

Scientific Explanation

Model

Scale Model

Prediction

Scientist

Three–Dimensional (3-D)

Two–Dimensional (2-D)

Space Science Vocabulary

Atmosphere

Air Resistance

Force

Mass

Gravity

Weightless

Satellite

Orbit

Diameter

Sphere

System

6. With the wide-tip pen, write the nine key concepts that follow on sentence strips. Have them handy to post on the concept wall under *What We Have Learned About Space Science* during the session.

Gravity is an invisible pulling force.
All objects have gravity that pulls on all other objects.
Objects with more mass have more gravity.
The farther apart objects are, the less the pull of gravity between them.
Earth's gravity pulls things toward the center of the Earth.
It is hard to leave Earth because of the pull of Earth's gravity.
Gravity can pull across great distances.
People have been gathering evidence about space for a long time.
Current missions continue to provide evidence about space.

Introducing Gravity

1. Ask how to make a pen stick to a book. Hold up a pen or pencil and a book, side by side and separated by at least several hand widths. Be sure to hold the book vertically as you ask the question, "How could I get this pen to stick to the book?" If students make suggestions such as "You could stick it to the book with glue," say, "Yes, but how could I make it stick without glue?"

Unit Goals

Earth, like all planets and stars, is round like a sphere.

Earth's gravity pulls things toward the center of Earth.

Gravity is everywhere.

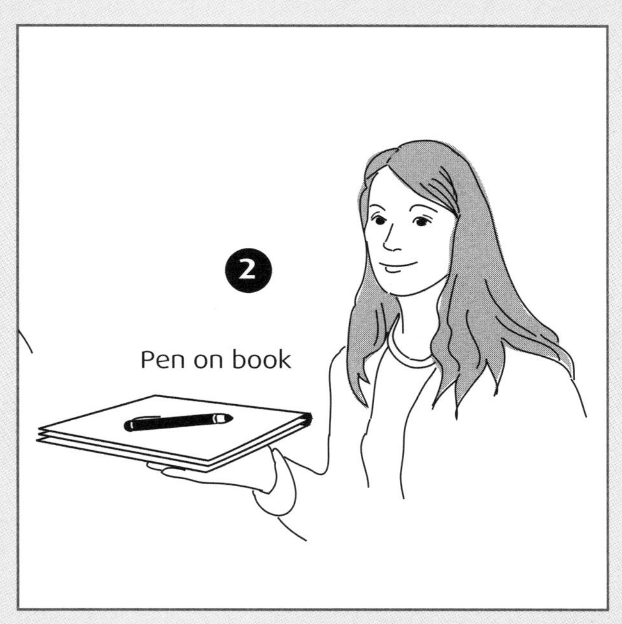

Key Vocabulary

Science and Inquiry Vocabulary

Evidence

Scientific Explanation

Model

Scale Model

Prediction

Scientist

Three–Dimensional (3-D)

Two–Dimensional (2-D)

Space Science Vocabulary

Atmosphere

Air Resistance

Force

Mass

Gravity

Weightless

Satellite

Orbit

Diameter

Sphere

System

2. Use an invisible force. Listen to a few responses, and then tell them that you can make the pen stick to the book using an *invisible force.* Turn the book so that it is horizontal, and drop the pen on the book. Point out that the pen is sticking to the book.

3. Getting a pen to stick to table or floor. Now, ask them to suggest how you could get the pen to stick to a table, using the same invisible force. [Set it on the table.] Ask how you could get it to stick to the floor, and so on.

4. Gravity is a pulling force. Ask, "What is the invisible force that pulled the pen and book together?" [*Gravity.*] Tell them that *gravity is an invisible pulling force between objects.*

5. What have you heard about gravity? Ask students what they know or have heard about gravity. If they make inaccurate statements, do not correct them at this point. Take note of their ideas so that you can address them later on.

6. Every object that has mass exerts gravitational pull. Say that the book and the pen have *mass,* which means that they are made of *matter,* or "stuff." Every object with mass has a gravitational pull between it and other objects, including the Sun, Moon, Earth, the pen, the book, their bicycles, their houses, and even they themselves.

7. Two things affect how strong gravity is. Say that the amount of gravity pulling two masses together depends on two things:

#1 *How much mass the objects have*. Things with more mass exert more gravitational pull.

#2 *How far apart the objects are*. The farther apart objects are, the less the pull of gravity between them. (It gets weaker as objects get farther away from each other, but it never stops working: gravity pulls over infinite distances.)

8. Earth has great mass and it's very close to us. Say that the Earth has much more mass than pencils, bicycles, people, or houses. Because Earth is so massive and close to us, its gravity has a strong effect on us, and everything around us. Mention that it was really mostly the gravitational pull between the Earth and the pen that kept the pen on the book: the book has some mass and gravitational pull, but it's insignificant compared with the Earth's gravity.

Unit Goals

Earth, like all planets and stars, is round like a sphere.

Earth's gravity pulls things toward the center of Earth.

Gravity is everywhere.

TEACHER CONSIDERATIONS

SCIENCE NOTES

Gravity is a force that operates between objects, such as the Earth and a pair of scissors, or the Earth and the Moon. Phrasing a statement about gravity as "There is gravitational pull between the Earth and a pair of scissors" is more accurate than saying, "The Earth's gravity pulls on a pair of scissors." See *Background*, page 32–34.

TEACHING NOTES

You may want to point out that neither thoughts nor ideas have mass, so they don't have gravity.

What One Teacher Said

"They had many questions to ask and were very eager to share their comments, which were always relevant to the topic at hand. They really wanted to know how gravity worked! It was challenging for some of the 4th graders to grasp the concept that "down" is always towards the center of the Earth, but I think that most of them "got it" by the end."

Key Vocabulary

Science and Inquiry Vocabulary

Evidence

Scientific Explanation

Model

Scale Model

Prediction

Scientist

Three–Dimensional (3-D)

Two–Dimensional (2-D)

Space Science Vocabulary

Atmosphere

Air Resistance

Force

Mass

Gravity

Weightless

Satellite

Orbit

Diameter

Sphere

System

PEOPLE AROUND THE WORLD

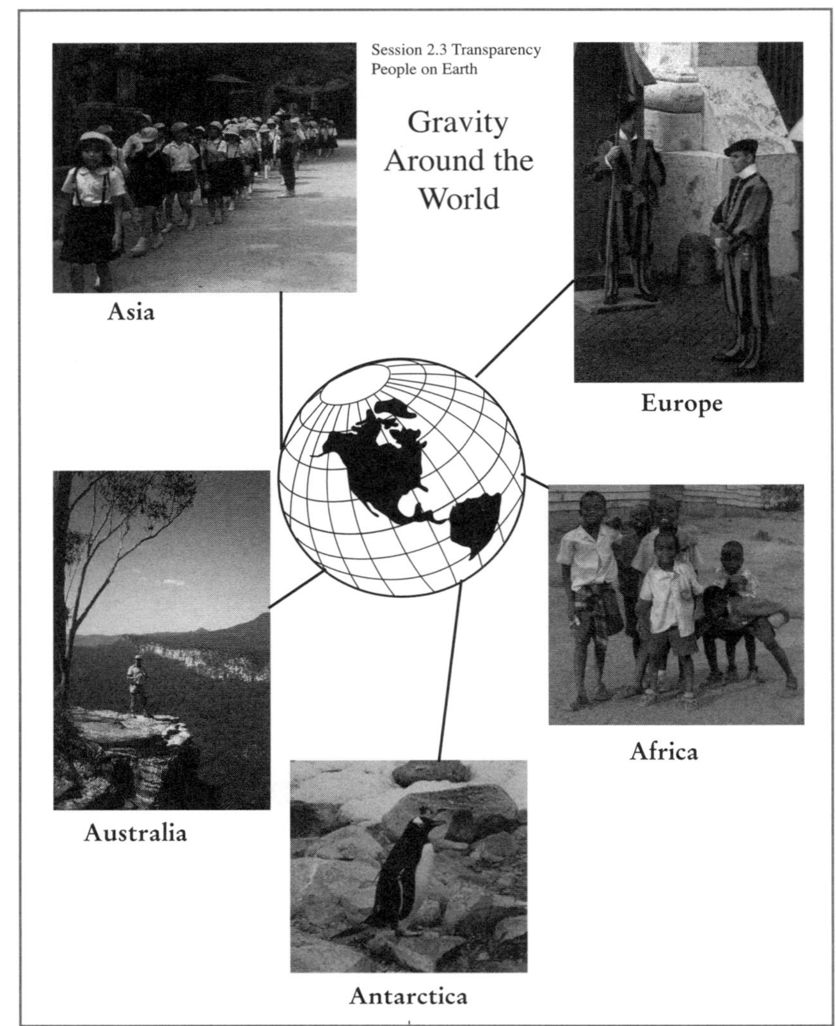

Revisiting Questionnaire Question #4

1. Gravity on different parts of the Earth. Ask if any students have been to a different part of the world, especially in the Southern Hemisphere. If so, ask how gravity worked there. Were they standing upside-down? If they dropped something, did it fall toward the sky, to the side, or toward the ground?

2. Photos of people in different places on Earth. Show the transparency or CD-ROM file of *People Around the World*. Review by asking, "Why does the Earth look flat in each photo?" [You are seeing only a very small part of the Earth.] Point out that all of the people pictured are standing up, with their feet on the ground. Emphasize that everywhere on Earth, the pulling force of *Earth's gravity pulls objects toward the center of the Earth.* "Down" means "toward the Earth," no matter where you live on Earth.

3. Discuss Question #5. On the board, sketch the Earth with people on different sides of it, as on the questionnaire. Show how each rock would fall by drawing lines from each person's hand straight toward the center of the Earth, landing on the surface.

Unit Goals

Earth, like all planets and stars, is round like a sphere.

Earth's gravity pulls things toward the center of Earth.

Gravity is everywhere.

TEACHER CONSIDERATIONS

QUESTIONNAIRE CONNECTION

The activities in this session deal directly with the information relating to Question #5 on the *Pre-Unit 2 Questionnaire*.

> 5. This drawing shows some enlarged people dropping rocks at various places around the Earth. What happens to each rock? Draw a line showing the complete path of the rock from the person's hand to where it finally stops. Why will the rock fall that way?

Question #5: If you noticed in discussions or in scoring the questionnaires that your students are struggling with question #5, you may choose to spend more time on the discussion of this question now. Ask students to come to the board and show how each rock would fall, and draw and explain their reasoning.

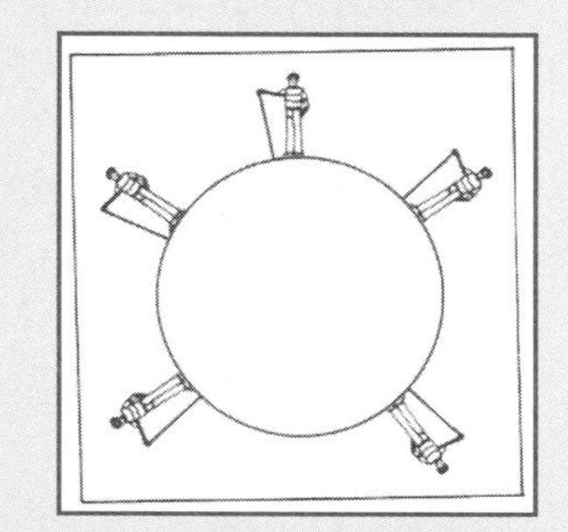

ASSESSMENT OPPORTUNITY

Critical Juncture—Where Is Down?

If your students show some resistance to the idea that "down" is always toward the center of the Earth, try one or both of the activities described in *Providing More Experience,* below.

PROVIDING MORE EXPERIENCE

1. Paper Dolls on a Globe. Tape three or four paper dolls to different positions on a globe. Say that they represent real people standing in different places on Earth. For each of the paper dolls, ask students to explain:

a. Which way would a rock fall if this paper doll dropped it? Why?

b. Which way is "down" for this paper doll? [Toward their feet.]

c. Which way is the sky for each paper doll? [Above their heads.]

d. Is "up" the same direction for all these paper dolls? [No, if you drew an arrow to show "up" for each doll, all the arrows would go in different directions. Wherever you are standing on Earth, "up" is simply away from the Earth.]

e. Does the doll standing in Australia feel as though they are hanging upside down? [No, they feel like we do, that "down" is toward their feet.]

Continued on page 279

Key Vocabulary

Science and Inquiry Vocabulary

Evidence
Scientific Explanation
Model
Scale Model
Prediction
Scientist
Three–Dimensional (3-D)
Two–Dimensional (2-D)

Space Science Vocabulary

Atmosphere
Air Resistance
Force
Mass
Gravity
Weightless
Satellite
Orbit
Diameter
Sphere
System

Beating Gravity

1. Ask if they think that they can beat the gravitational pull between themselves and Earth. Challenge them to demonstrate ways to "beat" the pull of gravity between each of them and the Earth. [They might jump up, lift an object, stand up, and so on.] With each demonstration, congratulate them on *appearing* to beat gravity. Point out that when they jump, they always come back to Earth. Ask, "Can you jump off the planet?" Tell them that's what they'd have to do to *really* beat the pull of gravity between them and the Earth.

2. Jumping 25,000 mph (4.3 km per second) to escape gravitational pull between you and Earth. Say that to jump off the planet, they'd have to jump at about 25,000 miles per hour (4.3 kilometers per second). Of course, there is no way that a person can jump anywhere near that fast. And even if we were able to jump at *24,000 mph*, we would get very high, but would eventually slow down and be pulled back toward Earth. It takes very powerful rockets to go 25,000 mph, get off the planet, and not fall back.

Unit Goals

Earth, like all planets and stars, is round like a sphere.

Earth's gravity pulls things toward the center of Earth.

Gravity is everywhere.

PROVIDING MORE EXPERIENCE, *CONTINUED*

2. A Hole Through the Center of the Earth

If you have time, and you think your students are ready for a more challenging question about gravity, make a sketch like the one at the right. Give students 10–15 minutes to think about it individually and discuss it in small groups. Then, discuss it in the large group. Draw several circles on the board, showing a stick figure dropping a rock in the tunnel in each one. Invite students to come up and draw their answers until several different ideas are represented. Then, lead a discussion debating the merits of each idea.

Don't reveal the "answer" to them until at least a day has passed. Tell them that this is a problem that stumps many adults! The best way to explain what occurs is to explain the history of the concept of gravity in this way:

Aristotle probably would have drawn a line to the center of the Earth: When the ancient Greeks came up with the idea of a ball-shaped Earth, they had to explain why people who lived on the other side of the world didn't fall off. Aristotle, who lived about 2,300 years ago, thought that everything went to its "natural resting place" in the center of the Universe, which he thought was at the center of the Earth. If Aristotle had to answer this question, he probably would have drawn a line to the center and stopped there.

Newton Named Pulling Force "Gravity." About 300 years ago, Isaac Newton said that the rock falls because of a pulling force between every particle within the Earth and every particle within the rock. He named the force gravity.

According to Newton, from the rock's point of view, "down" is always toward the greater mass of the Earth. Before it reaches the center of the Earth, the rock goes faster and faster because it is still falling "down." It starts slowing only after it passes the center, because then the greater mass of the Earth is behind it.

If Isaac Newton were to answer this question, he probably would draw the rock falling back and forth between the two poles of the Earth, until air resistance finally slowed it down. Eventually, it would settle in the exact center of the Earth, suspended in the middle of the tunnel.

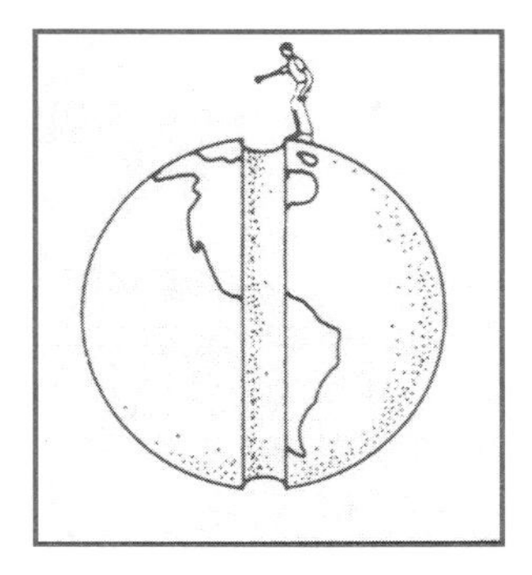

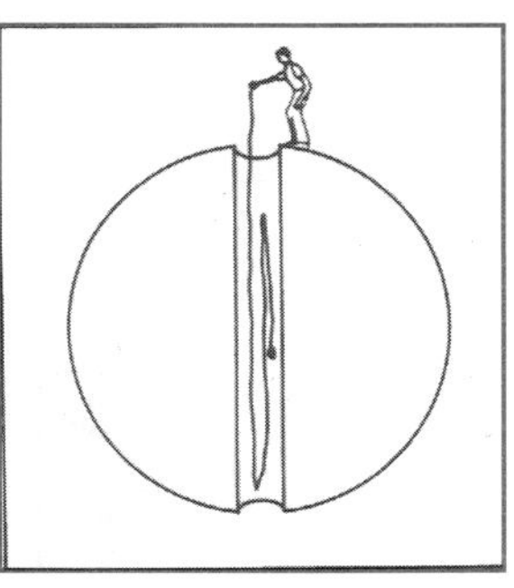

Key Vocabulary

Science and Inquiry Vocabulary

Evidence
Scientific Explanation
Model
Scale Model
Prediction
Scientist
Three–Dimensional (3-D)
Two–Dimensional (2-D)

Space Science Vocabulary

Atmosphere
Air Resistance
Force
Mass
Gravity
Weightless
Satellite
Orbit
Diameter
Sphere
System

Small-Group Activity: Is Gravity Strong or Weak?

1. Introduce evidence circle activity. Assign your students to evidence circles of four. Tell students to discuss the question, "Is gravity strong or weak?" Tell them that one goal of the discussion is to get them to think about and discuss gravity's effects on Earth and in space. Another goal is for them to practice using evidence to back up their thinking. Explain that the procedure will be a little different than it was in Session 2.2:

Evidence Circle Procedure

a. One student says what he/she thinks and the reasons why.

b. The students who agree add their reasons.

c. Then, the students who disagree say why, and present their reasons.

d. The team discusses the question with one another to see if they can come to an agreement. If no one disagrees, they can talk about all the evidence that makes them all convinced of their view.

2. Use evidence. Emphasize that when it is their turn, each student should give evidence to support why they think gravity is strong or weak. Ask, "What is one example of evidence that you could use to back up the statement that gravity is weak?" [They can beat gravity by jumping.] Ask for an example of evidence that "gravity is strong." [We can't jump off the Earth.]

3. *Optional:* Record the evidence. If time allows, have students record all the evidence they generate in their evidence circles. (Please see right hand page, 281)

Unit Goals

Earth, like all planets and stars, is round like a sphere.

Earth's gravity pulls things toward the center of Earth.

Gravity is everywhere.

TEACHER CONSIDERATIONS

TEACHING NOTES

Procedure for evidence circles: This activity uses the same procedure introduced in Unit 1. If your students experienced Unit 1, just remind them of the procedure. If not, introduce it now. You might want to post the four steps on the board or on a chart.

PROVIDING MORE EXPERIENCE

Record the evidence in evidence circles. If you have time, have students list the evidence they generate during their evidence circle discussions.

1. Write down the evidence for both sides. Say that as students give evidence for "strong" or "weak," everyone on their team will write down the evidence. Put the following headings on the board:

Is Gravity Strong or Weak?

Evidence That Gravity Is Strong:	Evidence That Gravity Is Weak:
____________________	____________________
____________________	____________________
____________________	____________________
____________________	____________________
____________________	____________________

2. Hand out a piece of lined paper to each student. Have them each fold a paper in half and write the headings on each column. Remind them of the evidence circle procedure, and have them begin.

What One Teacher Said

"The evidence circles were great. I can see the students beginning to cite good evidence for the difference in gravity on the moon vs. the earth."

Key Vocabulary

Science and Inquiry Vocabulary

Evidence
Scientific Explanation
Model
Scale Model
Prediction
Scientist
Three–Dimensional (3-D)
Two–Dimensional (2-D)

Space Science Vocabulary

Atmosphere
Air Resistance
Force
Mass
Gravity
Weightless
Satellite
Orbit
Diameter
Sphere
System

Large-Group Discussion of Gravity

1. Regain the attention of the class, and ask students to share evidence in support of the idea that gravity is weak or strong.

Evidence for gravity being weak might include:

- We can jump in the air, breaking the pull between ourselves and Earth.
- Spaceships beat the pull between them and Earth when they take off.
- On Earth, we can't feel the gravity from distant stars.
- Gravity can't stop us from throwing or lifting things.

Evidence for gravity being strong might include:

- Gravity affects almost everything we do in everyday life on Earth. Things on Earth and in space are affected by the gravitational pull between them and Earth.
- The gravitational pull between Earth and all the "stuff" on Earth holds everything on Earth, including air.
- There is gravitational pull between Earth and spaceships even when the spaceships are in space.
- The gravitational pull between the Earth and satellites and the International Space Station keeps them in orbit.
- Even though it is far away in space, the gravitational pull between the Earth and the Moon keeps the Moon in orbit around the Earth.
- Even though the Sun is very far from the Earth, the gravitational pull between the Sun and the Earth keeps the Earth in orbit around the Sun.

2. Weak in some ways, strong in others. Conclude by saying that gravity could be considered weak or strong, depending on how you look at it. We can "beat" gravity in some ways, but gravity can pull across great distances in space. Even though gravity gets weaker the farther away objects that are from each other, it goes on forever.

Unit Goals

Earth, like all planets and stars, is round like a sphere.

Earth's gravity pulls things toward the center of Earth.

Gravity is everywhere.

TEACHER CONSIDERATIONS

TEACHING NOTES

Tips for Leading Good Discussions: Engaging students in thoughtful discussions is a powerful way to enrich their learning. Even the most experienced teacher can use a few reminders about how to lead a good discussion. Listed below are some general strategies to keep in mind while leading a class discussion:

- Ask broad questions (questions which have many possible responses) to encourage participation.
- Use focused questions sparingly (questions which have only one correct response) to recall specific information.
- Use wait time (pause about 3 seconds after asking a question before calling on a student).
- Give non-judgmental responses, even to seemingly outlandish ideas.
- Listen to student responses respectfully, and ask what their evidence is for their explanations.
- Ask other students for alternative opinions or ideas.
- Try to create a safe, non-intimidating environment for discussion.
- Try to call on as many females as males.
- Try to include the whole group in the discussion.
- Offer "safe" questions to shy students.
- Employ hand raising or hand signals to insure whole group involvement.
- Take time to probe what students are thinking.
- Consider your role as a collaborator with the students, trying to figure things out together.
- Encourage students to figure things out for themselves, rather than telling them the answer.

Key Vocabulary

Science and Inquiry Vocabulary

Evidence

Scientific Explanation

Model

Scale Model

Prediction

Scientist

Three–Dimensional (3-D)

Two–Dimensional (2-D)

Space Science Vocabulary

Atmosphere

Air Resistance

Force

Mass

Gravity

Weightless

Satellite

Orbit

Diameter

Sphere

System

3. Post key ideas about gravity on the wall. Post on the concept wall the seven key concepts about gravity you prepared earlier. Read and review them with the students, and leave them up for the remainder of Unit 2.

Gravity is an invisible pulling force.
All objects have gravity that pulls on all other objects.
Objects with more mass have more gravity.
The farther apart objects are, the less the pull of gravity between them.
Earth's gravity pulls things toward the center of the Earth.
It is hard to leave Earth because of the pull of Earth's gravity.
Gravity can pull across great distances.

Reading: Yuri Gagarin – First Person in Space

1. Reading about the first person in space. Tell the class that they're going to read about the first human being ever to go on a mission to orbit the Earth in space. The rocket that took his spacecraft up into space went fast enough to let him "beat" gravity and leave Earth's atmosphere. Tell students that one goal of the reading is to help them understand the concepts about the Earth's shape and gravity which they have been learning about.

2. Review *atmosphere* and *orbit*. Ask, "Is there air in space?" [No.] "Where does space begin?" [Earth's atmosphere gets thinner gradually, the higher you go. It's hard to say exactly where space begins, but you could say it begins about 500 km above the Earth.] "What does it mean to *orbit* Earth?" [Travel around it in space, under the influence of Earth's gravity.]

Unit Goals

Earth, like all planets and stars, is round like a sphere.

Earth's gravity pulls things toward the center of Earth.

Gravity is everywhere.

TEACHER CONSIDERATIONS

TEACHING NOTES

Purposes of the Reading: *Yuri Gagarin – First Person in Space*

- Provides evidence of the spherical Earth.
- Helps develop understanding that gravity keeps objects in orbit. (The understanding that satellites and the Moon orbit the Earth, and that the Earth orbits the Sun come into play throughout the unit.)
- Reinforces the idea that there is air in the atmosphere, which allowed Yuri to parachute back to Earth in his spacecraft.
- Reintroduces the concept of the continuing story of space exploration—of people sending spacecraft and themselves up into the sky to explore and seek evidence.
- Revisits measurement and scale: size of spaceship; altitude of flight; and altitude of atmosphere.
- Helps introduce the idea that there is gravity in space, even without air. Many students think that gravity is caused by air, so they mistakenly think that there is no gravity in space, on the Moon, or on other planets.
- Introduces example of the feeling of weightlessness, despite the presence of gravity.

Define *atmosphere* before the reading. In Unit 1, Session 1.2, students were introduced to the Earth's atmosphere with the story of the sheep, duck, and chicken balloon trip. If your students did not learn about the atmosphere previously, tell them that Earth has an *atmosphere*, or layer of air around it. Now that they have a better understanding of the spherical Earth, they can picture the atmosphere as a relatively thin blanket of air surrounding the planet.

Was Yuri Gagarin really in space? The altitude of 500 km was used in Unit 1 to define where space begins. Gagarin flew up to 327 km above the Earth. This is an area where there is still a bit of atmosphere, but the air is extremely thin. Although he was below 500 km, it is not disputed that Gagarin was the first person in space. The boundary between the atmosphere and space is gradual, so any exact definition is somewhat arbitrary.

Key Vocabulary

Science and Inquiry Vocabulary

Evidence

Scientific Explanation

Model

Scale Model

Prediction

Scientist

Three Dimensional (3-D)

Two Dimensional (2-D)

Space Science Vocabulary

Atmosphere

Air Resistance

Force

Mass

Gravity

Weightless

Satellite

Orbit

Diameter

Sphere

System

YURI GAGARIN, PAGE 1

Name:______________________________

Yuri Gagarin: The First Person in Space

Yuri Gagarin grew up on a farm in Russia. Even when he was very young, he knew he wanted to fly. Many years later Yuri became a pilot, but he still wanted to go higher. When Yuri heard about space travel, he asked to become an astronaut before anyone else in Russia. Many others wanted to be astronauts too, but after lots of tests and training Yuri was chosen.

Before Yuri, dogs and monkeys had been sent into space, so scientists knew animals could survive there. They were not sure if a person could work and think very well in space, though.

Yuri's spaceship was very small, but a huge rocket sent it into space. Some people said riding in the spaceship would be like sitting in a tin can on top of a bomb. As the rockets fired, Yuri shouted, "Let's go!" The rocket ship took off into the sky, and the crew on the ground felt nervous about Yuri. Finally they heard him say, "I see Earth. It's so beautiful!" Those were the first words any person ever said in space.

Yuri was the first person in space. He went to space on April 12, 1961. He was also the first person to orbit the Earth (travel around it in space). And he was the first person to see the Earth from space. He told people back on Earth that he could see Earth's shape—it really was round like a ball. He could also see the thin blue atmosphere around the Earth with black space above it.

As his spaceship orbited above the Earth, Yuri felt weightless. He floated in his spaceship. To gather evidence about what people can do while feeling weightless, Yuri ate food, drank water, and wrote notes. He found out that all those things are possible in space.

Yuri's spaceship traveled at a speed of about 3 kilometers per second (about 18,000 miles an hour). At that speed it only took him 108 minutes for one orbit around the Earth. When his spaceship slowed and came back down to Earth, Yuri slowly stopped floating and sank back in his seat. He was so happy he started singing. When his spaceship was in the atmosphere, but still very high above Earth, a parachute opened to slow the spaceship down. Yuri landed safely.

3. Review challenging words. Depending on the reading level of your students, review some or all of these potentially difficult words before the reading: *Yuri Gagarin, weightless, nervous, astronaut.*

4. Conduct a brief discussion before the reading. Ask some of the following questions to stimulate students' thinking:

a. What might be dangerous in a space mission? What kinds of things would scientists worry about when sending the first person on a mission into space?

b. What do you think it would feel like to be in a spacecraft
- waiting for takeoff?
- taking off?
- orbiting?
- landing on Earth?

5. Read *Yuri Gagarin—The First Person in Space*. Depending on students' reading abilities, you may want to have them do independent, paired, or shared reading.

6. Early finishers read second page. Say that if they finish reading the first page, they should continue and read the second page.

Unit Goals

Earth, like all planets and stars, is round like a sphere.

Earth's gravity pulls things toward the center of Earth.

Gravity is everywhere.

TEACHER CONSIDERATIONS

YURI GAGARIN, PAGE 2

TEACHING NOTES

If time is short: If time is short for the class discussion after the reading, you might want to use some of the discussion questions in a writing assignment.

Optional: If your class made a chart in Unit 1 of the distances to sky objects, you might measure and mark the altitude of Gagarin's flight on the chart. You could then compare it with the heights of other objects, such as the altitude of the International Space Station.

Other Space Missions: Your students may become curious about other space missions. See page 339 and the *Resources* section, pages 57-59.

Previewing the Kepler Mission: The NASA Kepler Mission searches for planets around other stars.

If you think that your students would be interested in hearing about this mission, you may want to describe to them how it relates to gravity. Tell students that gravity keeps Earth's air from flying off into space. We need this air to breathe. Tell them that some planets have air, and some do not. *How much gravity a space object has affects whether or not it can hold onto an atmosphere of air.* The amount of gravity a space object has is determined by its mass. Small planets with less mass, such as Mercury and Pluto, and most moons do not have air. Planets with more mass, like Jupiter and Neptune, have lots of air, though their air is very different from the air on Earth. Explain that some NASA scientists who are part of a mission called Kepler are going to search for planets around other stars—especially planets that might have life. They know that their search is a little like the story of Goldilocks: the biggest planets have way too much air to be good for life, and the smallest planets have *no* air, which is definitely not good for life. So the best planets will be not too big and not too little, but "just right."

Students can learn more about the Kepler Mission in Unit 4 of the Grades 6–8 *Space Science Sequence.*

Name:________________________

More About Yuri Gagarin

Yuri Gagarin was from Russia. In Russia, astronauts are called cosmonauts. Yuri was a metal worker before he became a pilot. He was inside his space capsule as it landed by parachute. On later flights, cosmonauts landed by jumping out of the capsule and parachuting to the ground. Yuri became very famous after his flight. He said that speaking to large groups made him more nervous than flying into space. He died in 1968 in a plane testing accident.

Yuri with his daughter

1962 1998

The First American in Space: Alan Shepherd became the first American in space when he made a 15-minute flight on May 5, 1961. He also walked on the moon in 1971, but he was not the first astronaut to travel there.

The First American to Orbit the Earth: On February 20, 1962, John Glenn became the first American to orbit the Earth. He orbited the Earth three times during a flight of about five hours. Later he became a U.S. Senator. In 1998 he went back into space for a nine-day mission on the space shuttle. On that mission, he became the oldest person to fly into space.

The First Woman in Space: Valentina V. Tereshkova, a Russian cosmonaut, was the first woman in space. On June 16, 1963, she orbited the Earth 48 times, which was a record at the time. Valentina had a lot of parachuting experience. When she returned to the atmosphere where there was air, she jumped out of her spacecraft and parachuted back to Earth.

The First American Woman in Space: Sally Ride became the first American woman in space when she spent eight days on the space shuttle in 1983. Before deciding to become an astronaut and a scientist, Sally thought about becoming a professional tennis player.

The First African-American in Space: Guion "Guy" Bluford became the first African-American in space when he flew on the space shuttle on August 30, 1983. He is also a pilot and a scientist. He later made three more trips into space.

Discussing the Reading

1. Bring out concepts of gravity and Earth's shape. Ask some questions about the story relating to the Earth's shape and gravity:

a. Why did Yuri need a big rocket? [To "beat" gravity and go into space.]

Note: Because he remained in Earth's orbit, he didn't completely "beat" the gravitational pull between his spaceship and Earth.

b. Did he fall right back down? [No, he orbited the Earth once.]

c. What shape did the Earth look like from Yuri's spaceship? [Round like a ball; spherical.]

d. Was there gravity in space? [Yes!] Emphasize that Earth's gravity reaches far into space. The reason that astronauts in orbit have the feeling of weightlessness has to do with moving fast around the Earth in orbit.
(Please see page 34 for more information about what causes the feeling of weightlessness in orbit.)

2. Post two more key concepts. Post the last two key concepts. Conclude the session by saying that Yuri's flight was only the beginning of space exploration by humans. If there is time, discuss some of the other space pioneers highlighted on the second page of the reading.

People have been gathering evidence about space for a long time.
Current missions continue to provide evidence about space.

Unit Goals

Earth, like all planets and stars, is round like a sphere.

Earth's gravity pulls things toward the center of Earth.

Gravity is everywhere.

TEACHING NOTES

Decide when to explain weightlessness in orbit: The important point to make in this session is that even though Yuri Gagarin felt weightless, there *was* a gravitational pull between his spaceship and the Earth. You may want to leave it at that, and tell students that they will find out more about why he felt weightless in the next session.

However, you may want to discuss the ideas below now, to help students begin grappling with the idea that an orbit is really a "free fall" around the Earth.

More information on this subject can also be found in the *Background for Teachers* section, page 34.

PROVIDING MORE EXPERIENCE

Why Did Yuri Feel Weightless?

1. Use the idea of a ladder to show that there is gravity in orbit. Explain that if a person could climb a 327-kilometer ladder up as high as Yuri's spaceship orbited, the person would still be able to stand on the ladder. The person would not feel weightless. They would feel the pull of gravity between Earth and their body almost as strongly as if they were standing on the ground. The reason that Yuri felt weightless in his spaceship was not because there was no gravity, or even because there was less gravity. He felt weightless because he was moving fast in an orbit around the Earth.

2. An orbit is a "free fall" around the Earth. Ask, "Why would moving in orbit make Yuri feel weightless?" Accept any ideas that students have. Tell students that scientists say that astronauts in orbit like Yuri, are in "free fall" around the Earth. This falling is why they feel weightless, not because there is no gravity.

Explain that scientists chose the exact speed so that Yuri's space ship would reach a certain height, fall sideways, and then keep falling in an orbit all the way around the Earth! This is what orbiting the Earth really is: the spaceship is falling around the Earth. Gravity is pulling the object toward Earth, but the speed of the object makes it continually "miss" the Earth, so it does not fall to the ground. If the object slows down, then it eventually does fall to Earth.

Key Vocabulary

Science and Inquiry Vocabulary

Evidence
Scientific Explanation
Model
Scale Model
Prediction
Scientist
Three–Dimensional (3-D)
Two–Dimensional (2-D)

Space Science Vocabulary

Atmosphere
Air Resistance
Force
Mass
Gravity
Weightless
Satellite
Orbit
Diameter
Sphere
System

Unit Goals

Earth, like all planets and stars, is round like a sphere.

Earth's gravity pulls things toward the center of Earth.

Gravity is everywhere.

TEACHER CONSIDERATIONS

OPTIONAL PROMPTS FOR WRITING OR DISCUSSION

You may want to choose one or more of the prompts below for science journal writing in class or as homework. The prompts could also be used for a discussion or during a final student sharing circle.

1. I didn't know this about gravity before:__________.
2. I'm still wondering this about gravity:__________.
3. List 10 ways that gravity makes things easier for you in your daily life (example: milk staying in an open glass), and ten ways it makes things harder (example: lifting a heavy pack).
4. List things that might be hard to do if you were weightless in a spaceship.
5. List things that you think would be easier to do living on a planet with less gravity.
6. Put these objects in order of which you think has most gravitational pull and which has least: a pencil, the Moon, a mountain, a car, cotton candy, Earth, an air-filled balloon, the Sun.

Key Vocabulary

Science and Inquiry Vocabulary

Evidence

Scientific Explanation

Model

Scale Model

Prediction

Scientist

Three–Dimensional (3-D)

Two–Dimensional (2-D)

Space Science Vocabulary

Atmosphere

Air Resistance

Force

Mass

Gravity

Weightless

Satellite

Orbit

Diameter

Sphere

System

SESSION 2.4

Weightlessness

Overview

The session begins with an introduction to a spring scale as a device used to measure the gravitational pull between an object and the Earth.

Student pairs rotate through a series of learning stations featuring drawings of people in different situations. For each situation, students discuss whether or not they think a person might feel weightless, and whether or not there is gravity. Students record their votes at each station. In a follow-up class discussion, students learn that, although a person might feel weightless in all of these situations, gravity is present in every one of them. The students attempt to explain the causes of the feeling of weightlessness.

Next, the class reads *The Vomit Comet*, which is about an airplane of that nickname that is used to create temporary situations for astronauts and others to experience weightlessness. The reading explains the causes of weightlessness and reinforces the concept that gravity exists everywhere in the Universe.

In a follow-up demonstration, a student weighs an object with a spring scale. The class carefully observes and records what happens to the object's weight measurement as the student jumps and lands with the spring scale in hand. They discuss how the spring scale shows temporary weightlessness, even though the gravitational pull between the object and the Earth remains in effect.

Finally, the class enjoys footage of people experiencing weightlessness in the Vomit Comet and in spacecraft. As they watch the footage, students are reminded that the gravitational pull between the people and other objects still exists, despite the fact that they are feeling weightless.

Weightlessness	**Estimated Time**
Introducing the feeling of weightlessness	5 minutes
Weightlessness stations and discussion	20 minutes
Reading and discussing *The Vomit Comet*	15 minutes
Measuring weightlessness and putting it all together	20 minutes
TOTAL	**60 minutes**

Unit Goals

Earth, like all planets and stars, is round like a sphere.

Earth's gravity pulls things toward the center of Earth.

Gravity is everywhere.

Key Vocabulary

Science and Inquiry Vocabulary

Evidence

Scientific Explanation

Model

Scale Model

Prediction

Scientist

Three–Dimensional (3-D)

Two–Dimensional (2-D)

Space Science Vocabulary

Atmosphere

Air Resistance

Force

Mass

Gravity

Weightless

Satellite

Orbit

Diameter

Sphere

System

What You Need

For the class

- ❑ overhead projector or computer with large screen monitor/LCD projector
- ❑ 1 spring scale
- ❑ 1 brightly colored toothpick
- ❑ small piece of poster tack
- ❑ 1 small binder clip (or other lightweight object)
- ❑ 2 large binder clips (or other objects)
- ❑ 1 copy of each of the 8 station sheets from the student sheet packet:
    1. On a swing
    2. In a spaceship flying through space
    3. On a roller coaster
    4. In an elevator
    5. Skydiving
    6. In an airplane flying up and down
    7. In a spaceship orbiting Earth
    8. In a drop cage at an amusement park
- ❑ 1 colored overhead transparency pen
- ❑ overhead transparency or CD-ROM file of the *Measuring Weightlessness* sheet from the transparency packet
- ❑ CD-ROM or video footage of weightlessness

For each student

- ❑ 1 copy of the reading *The Vomit Comet* from the student sheet packet

Getting Ready

The day before the activity

1. If you're not using poster tack, attach the brightly colored toothpick in a horizontal position on the pointer of the spring scale using glue or tape. This allows the class to see the pointer move during demonstrations in this session. The toothpick sticks best with poster tack, which doesn't take the extra time to dry and allows the toothpick to be reattached immediately if it falls off.

The day of the activity

1. Have materials handy. Have the spring scale, one small binder clip, and two large binder clips (or other light and heavier objects) near where you will do your demonstrations. Practice weighing the objects to be sure that their weights will work for the demonstrations at the beginning and end of the session.

Unit Goals

Earth, like all planets and stars, is round like a sphere.

Earth's gravity pulls things toward the center of Earth.

Gravity is everywhere.

TEACHER CONSIDERATIONS

TEACHING NOTES

Reading level: The reading level of page 1 of the reading is appropriate for most third and fourth graders. Page 2 and the additional pages on the CD-ROM are for students who are interested in further information on the topic, and who are able to read at a slightly higher level.

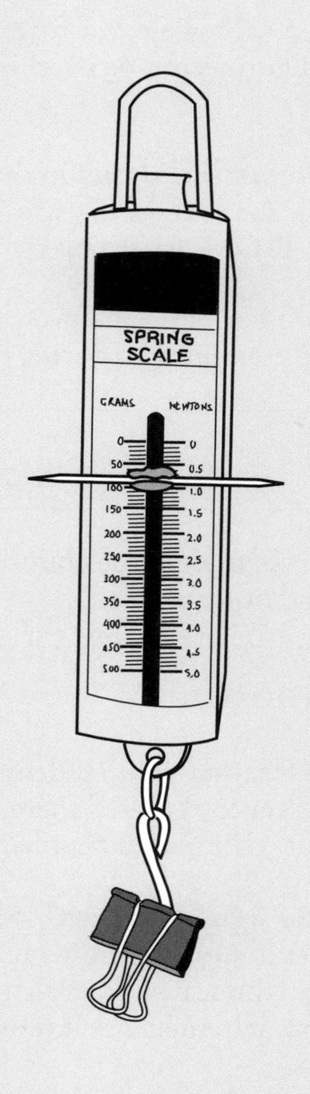

Key Vocabulary

Science and Inquiry Vocabulary

Evidence

Scientific Explanation

Model

Scale Model

Prediction

Scientist

Three Dimensional (3-D)

Two Dimensional (2-D)

Space Science Vocabulary

Atmosphere

Air Resistance

Force

Mass

Gravity

Weightless

Satellite

Orbit

Diameter

Sphere

System

2. **Make one copy of each of the eight weightlessness station sheets.** Decide on eight locations around the classroom where you will set the sheets out during the activity. They should be widely separated and on desks, tables, or counters, where they can be examined and discussed by up to four students at a time without crowding.

3. **Make one copy for each student of the reading *The Vomit Comet.***

4. **If you will not be using the CD-ROM, make an overhead transparency of *Measuring Weightlessness.***

5. **Arrange for the appropriate projector format** (computer with large screen monitor, LCD projector, or overhead projector) to display images to the class.

6. *Optional but highly recommended:* Set up the large screen monitor or LCD projector to show the footage of astronauts experiencing weightlessness on the CD-ROM. Alternatively, use a VCR and video.

7. **Write the following two key concepts on sentence strips,** and have them ready to post on the concept wall during the session.

Gravity is everywhere.
People feel weightless in some situations but there is still a pull of gravity between them and other objects.

Introducing Weightlessness

1. **Introduce the session.** Remind your students that Yuri Gagarin floated weightlessly in his spacecraft. Tell them that in today's session they will learn more about what makes people feel weightless.

2. **A spring scale measures weight on Earth.** Show your students the spring scale. Tell them that it measures how much something weighs on Earth. Show the 0 mark on the scale and explain how you attached the toothpick to the pointer to help them see it move.

3. **Weigh a small binder clip.** Attach a small binder clip (or other light object) to the scale, and show how the pointer/toothpick is pulled down to a number. That number is how much the object weighs. Say that when we measure the weight of an object using a spring scale, we are measuring the gravitational pull between that object and Earth.

Unit Goals

Earth, like all planets and stars, is round like a sphere.

Earth's gravity pulls things towards the center of Earth.

Gravity is everywhere.

TEACHER CONSIDERATIONS

CD-ROM NOTES

Weightlessness Videos: Imagine what it feels like to feel weightless in space by watching astronauts working and playing in space. Select videos of astronauts experiencing weightlessness by clicking on the video name at the bottom of the screen to begin watching a movie. To control the volume, video clip position, fast forward, rewind, pause, and restart, use the clear menu appearing at the bottom of the video. To start another video, click another mission name with your mouse. The videos were taken from the Apollo 10 mission, the Apollo 11 mission, the Skylab, the *Endeavour* shuttle, the *Discovery* shuttle, and the Microgravity University Program. Further instructions for using this program are included on the CD–ROM.

TEACHING NOTES

Keep the introduction short: Make your introduction and the demonstration with the spring scale as brief as possible in order to allow time for the other activities in the session. A second spring scale demonstration at the end of this session will more formally investigate how measured weight can change with motion.

Key Vocabulary

Science and Inquiry Vocabulary

Evidence
Scientific Explanation
Model
Scale Model
Prediction
Scientist
Three–Dimensional (3-D)
Two–Dimensional (2-D)

Space Science Vocabulary

Atmosphere
Air Resistance
Force
Mass
Gravity
Weightless
Satellite
Orbit
Diameter
Sphere
System

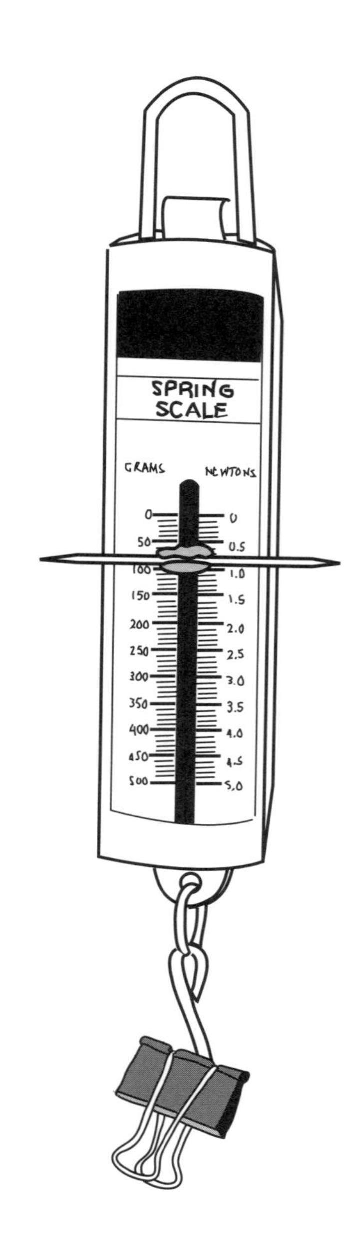

4. Weigh a large binder clip. Ask for predictions of what will happen if you weigh a larger binder clip. Remove the small binder clip, attach a large one, and let students see how the pointer is pulled down to a larger number. Tell them that objects that have more mass pull the spring scale down farther. There is more gravitational pull between the Earth and objects with more mass.

5. Ask a student's weight. Ask a student to volunteer his or her weight. Point out that weight is a measure of the gravitational pull between that student and the Earth. Tell the student that this measurement would be essentially the same if the student were standing anywhere on Earth.

6. Define weightlessness. Tell them that there are some situations in which objects or people may *seem* weightless even though they have mass. In these situations, people may temporarily feel as though they weigh nothing, and their weight would register as 0 on a scale. This is called *weightlessness*.

7. Have you ever seen anyone weightless? What do you think caused it? Ask if they can think of a situation in which they felt weightless. Take one or two examples from the students, and in each case, ask the class why they think a person might seem or feel weightless in that situation. Then, ask if anyone has a different explanation.

Students may bring up cartoons or martial arts movies as examples of weightlessness. Be sure to help them distinguish between these fictional situations and real situations.

Introducing the Weightlessness Stations

1. Introduce learning stations. Tell students that they will each go with a partner to some learning stations around the room. At each station, there is a sheet with a picture of a person in a different situation. Partners will discuss whether or not the person might feel weightless in that situation.

2. Hold up the *On a swing* station sign. Ask a student to help you model for the class how to discuss this question. Without concluding the discussion or giving too much away, discuss whether or not the person on the swing might feel weightless. Model for partners how to listen to each other's statements carefully.

Unit Goals

Earth, like all planets and stars, is round like a sphere.

Earth's gravity pulls things toward the center of Earth.

Gravity is everywhere.

TEACHER CONSIDERATIONS

SCIENCE NOTES

Spring Scale Measuring Units: In the *Measuring Weightlessness* activity with the spring scale, the focus is on noticing an increase or decrease in weight, rather than on specific measurements. This is partly because it is difficult for students to read a measurement on the spring scales from across the room. If you would like your students to know more about the units used in measuring gravitational pull, you may choose to introduce the units on your spring scale, which will likely be grams, Newtons, or pounds. These are different units that are used for measuring mass or weight on Earth. Pounds are used in only a few countries (including the United States). Pounds are not used by *scientists* anywhere in the world. Scientists use either metric units of mass (grams, kilograms) or *Systeme International* units of weight (Newtons).

The amount of pounds or Newtons something weighs would be different in places with different amounts of gravitational pull, such as on other planets. But grams are different. Grams and kilograms are units of mass. Mass, the amount of "stuff" something has, is a property that is constant, no matter how strong or weak the gravitational pull. You will have the same mass on the Earth, the Moon, on Jupiter, and everywhere in between. How many grams something is would be the same anywhere, no matter how strong or weak the gravitational pull. What makes this a bit confusing is that on Earth the exact same type of weighing device can be used to measure grams, Newtons, or pounds. So even though something would be the same number of grams on Earth as on the Moon, you couldn't use the same bathroom scale on the Moon as on Earth for an accurate measurement. You'd have to use a different device to measure grams on the Moon. For more on grams, Newtons, and pounds, see page 36 of the *Background for Teachers.*

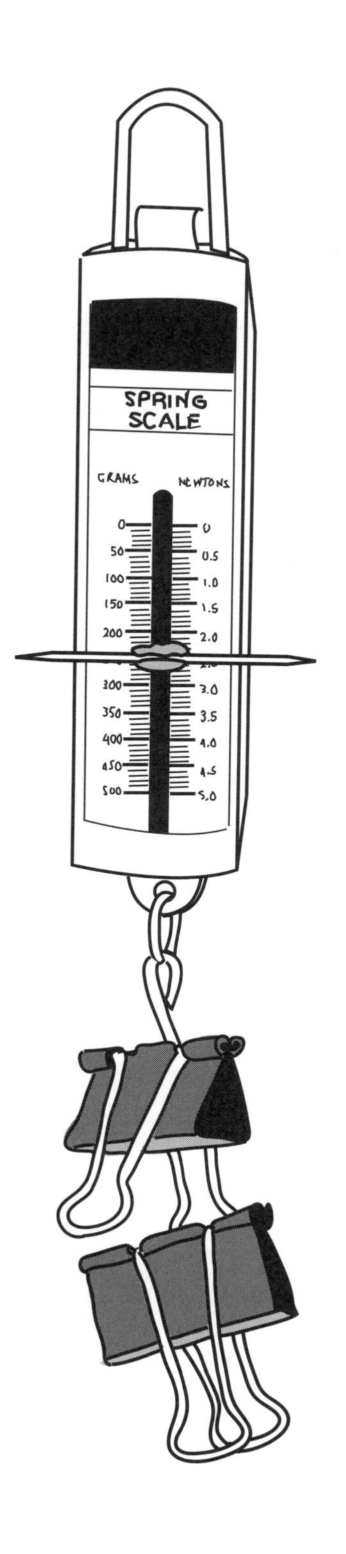

TEACHING NOTES

Keep students' attention during your introduction. When students find out that they will be working in partners, they are often so eager to know how they will be paired off that they don't hear your instructions about the activity. Tell students that you will make the partner assignments right before they get up for the activity, and ask them to focus on the instructions for now.

3. Show how to "vote" on the sheet. Point out the boxes on the left side of the sheet. Tell students that after discussing the question, each student will make either an "X" for YES or an "O" for NO, in any one of these squares. Demonstrate putting an "X" or "O" mark in the appropriate location.

4. The second question about gravity on the right side of the sheet. Explain that they will then discuss the second question on the sheet with their partner. Point out the question on the right side of the sheet: "Do you think that there is gravity here?" After they discuss this question, each student should mark an X for yes or an O for no in one of the boxes on the right side of the sheet.

5. Follow the same procedure at other stations. Tell students that they will move to a different station when they are ready, and follow the same procedure there. They do not need to go in any special order or wait for you to tell them to switch stations. They should choose stations that are not too crowded. It is okay for two pairs of students to be at a station at a time.

6. Don't rush. Emphasize that there is no reason to hurry: this is not a race. The point of the activity is to think about and discuss the questions. Tell students that there are eight stations, but they don't have to get to them all. Assign pairs of students to their first stations and have them begin.

Discussing the Weightlessness Stations

1. Tape station sheets to a wall and have students be seated. After each pair has had time to complete at least a few of the stations, tell them to stop and return to their seats with their partners. Tape the eight station signs on the wall in a row at the front of the classroom.

2. Discuss a few of the stations, including *Skydiving* and *In an airplane*.

a. Direct their attention to one of the station sheets. Point out how the class voted.
b. Ask a student who said they thought that a person could feel weightless there to explain why.
c. Ask if anyone has a different explanation.
d. Tell them that people *can* sometimes feel weightless in that situation.
e. In each case, tell them that there is gravity in that situation, even if people feel weightless there. (There is gravitational pull in all eight situations.)
f. Do this with only three or four of the station sheets, being sure to include the skydiving and airplane situations.

Unit Goals

Earth, like all planets and stars, is round like a sphere.

Earth's gravity pulls things toward the center of Earth.

Gravity is everywhere.

TEACHER CONSIDERATIONS

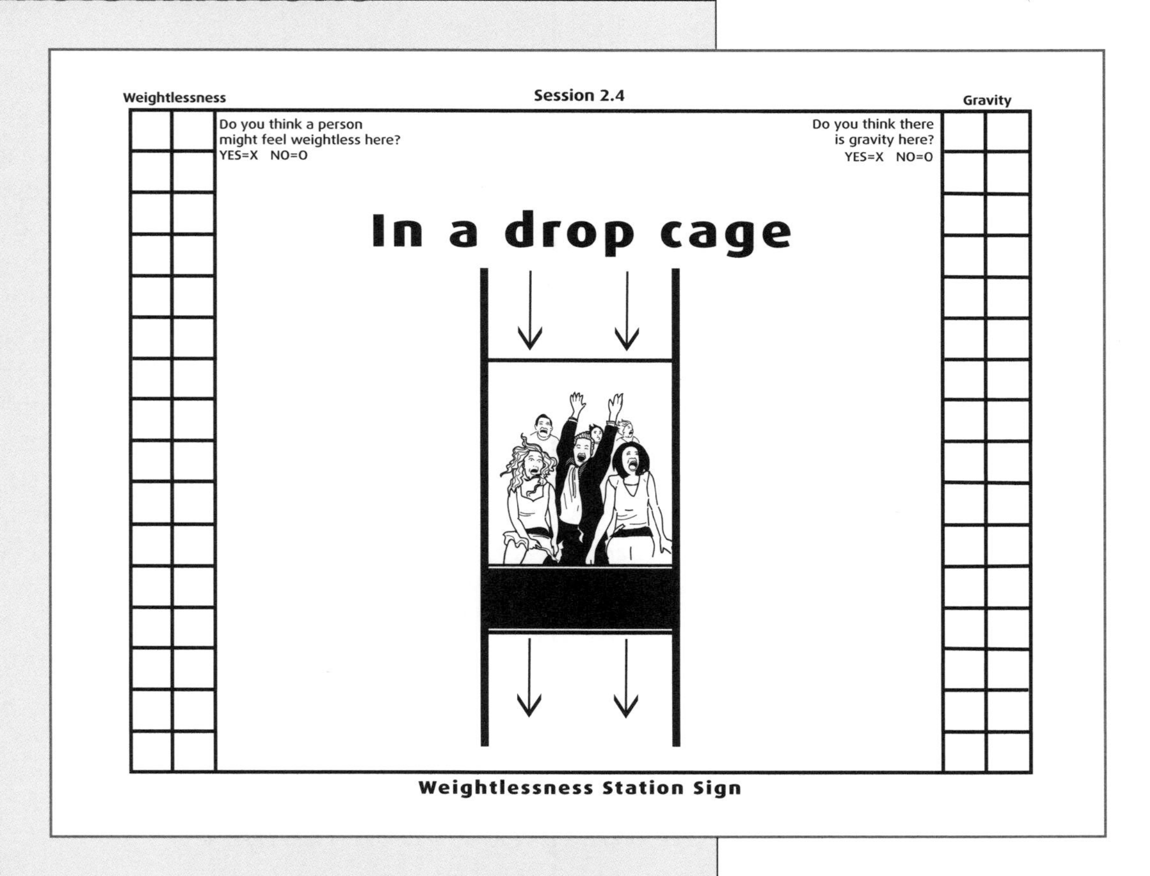

TEACHING NOTES

If students are not familiar with drop cages: Some of your students may not know what a drop cage is (station sheet #8), and would benefit from a brief explanation before the activity. Drop cages are amusement park rides in which participants are strapped inside a cage or elevator like contraption. It slowly rises up very high and then suddenly drops, giving the participants a brief free-fall experience. Different amusement parks have different names for these rides.

3. Discuss the skydiving station. As you discuss the skydiving station, point out that a person in that situation would feel weightless, but would also feel as though they were falling, because of the feel and noise of the air they'd be falling through, and because of watching the ground below getting closer. But imagine if a skydiver could be falling while inside of a large container that was also falling. They wouldn't feel or hear the air, and they wouldn't see the ground. Although they'd be falling, they would feel as though they were weightlessly floating.

4. Discuss the airplane flying up and down steeply. Tell them that this type of plane has been used to train astronauts. Being inside this plane when it dives down very steeply is similar to being in a large falling container. You are actually falling inside it while it is falling. You can feel as though you are weightlessly floating, even though you are falling.

5. Conclude that there is gravity in each situation. After discussing a few of the stations in this manner, tell them that a person could feel weightless in every situation shown, and that there is gravity in every situation.

Reading *The Vomit Comet*

1. Introduce the reading. Redirect their attention to the station with an airplane flying up and down very steeply. Say that they're going to read about a real plane that flies up and down very steeply. Officially, it's called the *Weightless Wonder,* but most people call it the *Vomit Comet* because some passengers feel sick in it. Say that this reading was written by a college student named Lauren who got to ride in the Vomit Comet.

2. Read *The Vomit Comet* as a group. Depending on their reading abilities, have students read independently or as a class. (A fascinating interview with more details is provided on additional pages on the CD-ROM.)

Unit Goals

Earth, like all planets and stars, is round like a sphere.

Earth's gravity pulls things toward the center of Earth.

Gravity is everywhere.

TEACHER CONSIDERATIONS

TEACHING NOTES

Purposes of the *Vomit Comet* Reading

- Reinforces the concept that there is gravity everywhere
- Helps students understand that gravity is present even though the motion of the airplane may induce a feeling of weightlessness
- Provides an example of how, throughout history, people have performed tests and experiments to find evidence to answer scientific questions
- Reinforces the idea of sky and space exploration—people sending aircraft, spacecraft, and themselves up into the sky on missions to explore and seek evidence

Key Vocabulary

Science and Inquiry Vocabulary

Evidence
Scientific Explanation
Model
Scale Model
Prediction
Scientist
Three–Dimensional (3-D)
Two–Dimensional (2-D)

Space Science Vocabulary

Atmosphere
Air Resistance
Force
Mass
Gravity
Weightless
Satellite
Orbit
Diameter
Sphere
System

THE VOMIT COMET, PAGE 1

Name:______________________________

The Vomit Comet

My name is Lauren, and I rode on the Vomit Comet. The Vomit Comet is the name some people use for a special kind of airplane. Riding in the airplane helps people find out what it's like to feel weightless. Many astronauts have trained in the Vomit Comet.

I was very excited when the plane took off. It took off like a normal plane, but then it flew up very steeply. It flew up so steeply that I felt very heavy. Even though the pull of gravity was the same, it felt like the pull of gravity was twice as strong as it is on Earth.

Then the plane started to fly down very steeply. That's when I felt weightless, and I floated in the air. It didn't feel like I was falling, but that's what I really was doing. I felt weightless for about 25 seconds.

Flying in the Vomit Comet is kind of like being on a swing. When you swing up high, before coming back down, you may feel weightless. Sometimes you may feel weightless riding down a very steep part of a roller coaster. The feeling may only last a second. But in the Vomit Comet it lasts 25 seconds. People in the Vomit Comet get to float around for 25 seconds at a time.

After 25 seconds, the plane went back up. Then I felt very heavy again. The plane flew up and down like that over and over again. It usually goes up and down 40 to 60 times in one flight. It's kind of like being inside a giant roller coaster with no seat belts. This makes a lot of people feel sick. That's why it's called the Vomit Comet.

Some people think there is no gravity when you feel weightless on the Vomit Comet. But you only feel weightless because you are falling inside the plane while the plane is flying down steeply. The invisible pull of gravity is everywhere. There is even gravity in space. Astronauts in orbit are actually falling all the time, without ever landing. They are not falling straight down towards the ground. Have you ever seen a car go "flying" off a cliff in a movie? The car is falling, but it's also moving sideways. Astronauts in orbit are falling, but they are so high and moving sideways so fast that they actually "fly" around the whole Earth and never land. That's what orbiting is. There is gravity everywhere, but there are places where a person can feel weightless. In these places it feels like there is no gravity. The Vomit Comet is one of those places.

Session 2.4 Student Reading, Page 2

Interview with Lauren

Question: Were you ever upside down?

Lauren: Yes, but I didn't feel like I was upside down. When you feel weightless, upside down and right side up feel exactly the same. I didn't feel like I was being pulled "down," even though I was actually falling down towards the Earth.

Question: Did you ever fall on your head?

Lauren: The crew on the Vomit Comet is very careful and makes sure no one gets hurt. Every time we changed from feeling weightless to feeling extra heavy, they would call out "feet down, coming out." We always had time to make sure we'd land on our feet and not on our heads.

Question: What experiment were you doing?

Lauren: I did two experiments on the Vomit Comet. One experiment tested different ways of putting medicines or nutrients directly into an astronaut's blood. Liquids move differently when they are weightless. We tested two different pumps to see how well they would work.

My other experiment was a model of how asteroids run into each other. I was interested in learning about how asteroids would spin after they were hit. I tested small models of asteroids made of water and sand. Evidence collected by spacecrafts near asteroids shows that some asteroids are made of very weak material like water and sand. On Earth, water and sand make a flat puddle, because the gravity on Earth feels so strong. On the Vomit Comet, I could make the water and sand stick together and float like a ball. When my water and sand ball was floating, I could hit it with another model asteroid and observe how it would spin.

Question: Did you vomit? Did anybody on your plane vomit? What happened to the vomit?

Lauren: I was lucky—I didn't vomit or feel sick at all. One boy who was on the plane with me did vomit. Everyone is given plastic bags to carry with them on the plane. We were all lucky that the boy

Discussing the Reading

1. Discuss the Reading. After the reading, ask the following questions, and allow as much discussion as possible:

a. When do people feel weightless in the plane, and when do they feel heavier? Why?

b. Is there anywhere in the Universe where there is no gravity? [No, there is gravity everywhere in the Universe.]

c. If there is gravity everywhere in the Universe, why do people sometimes feel weightless?

2. Though sometimes people feel weightless, there is gravity everywhere. Tell the class that many people think there is no gravity in space, and they think that that's why astronauts float around weightlessly in their spacecrafts. Emphasize that although gravity is stronger in some places than others, there is gravity everywhere. Astronauts feel weightless even though they are still pulled toward other objects by the force of gravity.

3. Add to the concept wall. Tell students that by considering different situations and reading the story, they have learned two key concepts about space science that are important to remember as they continue to study this subject. Post the sentence strips which you prepared earlier on the concept wall.

Gravity is everywhere.
People feel weightless in some situations but there is still a pull of gravity between them and other objects.

Unit Goals

Earth, like all planets and stars, is round like a sphere.

Earth's gravity pulls things toward the center of Earth.

Gravity is everywhere.

PROVIDING MORE EXPERIENCE

Additional discussion or writing prompts related to the reading: After the reading, you could use the additional discussion questions below to review some of the concepts. Alternatively, these prompts could be used for writing assignments:

a. What do you think it would feel like to be on the Vomit Comet?

b. If you were in a spaceship traveling between the Earth and Moon, do you think that you would feel your normal weight, heavier, or weightless during these situations:

- When the rocket first takes off from Earth
- When the rocket is coasting through space
- When the rocket is landing on the Moon
- When the rocket has finished landing on the Moon

c. What do you think might happen to each of these on the Vomit Comet:

- water
- a feather
- a large rock
- hair
- dust
- food
- vomit

THE VOMIT COMET, PAGE 2

who got sick vomited into his bag. If he had missed his bag, the vomit would have floated around when we felt weightless. But it would have fallen on the floor of the plane and made a mess when we felt heavy.

Session 2.4 Student Reading, CD-ROM Page 3

Question: Did your hair stand up?

Lauren: My hair went in every direction. There are so many things in everyday life affected by gravity that we don't even think about. On the Vomit Comet, things would float out of my pockets. Every time I tried to move just a little my whole body would float around.

Question: What would happen to water on the Vomit Comet?

Lauren: Water turns into round blobs and floats around while the Vomit Comet flies down. When the Vomit Comet flies up and feels very heavy the water falls to the ground.

Question: What is the Vomit Comet used for?

Lauren: It is mainly used for three things:

Practice. Lots of things are different when you feel weightless. Even something simple like turning a screw is very different. That's why it's good for astronauts to practice doing things while feeling weightless in the Vomit Comet before they try to do them in space.

Experiments. Sometimes people do experiments to see what will happen to things when they feel weightless. There have been experiments with bubbles, bees, and many other things.

Movies. Many parts of the movie Apollo 13 were filmed in the Vomit Comet.

Question: Can they make you feel like you're on the Moon or Mars on the Vomit Comet?

Lauren: Sometimes someone on the Vomit Comet shouts out "Martian-1" just as the plane begins to go back down. That means that it will feel like the gravity on Mars as the plane goes back down. The plane will not be flying down as steeply. If you stand up at Martian-1, you do not feel weightless. You feel lighter, like you would on Mars. Gravity on Mars is about two-thirds less than Earth's gravity. The plane can also make it feel like the gravity on the Moon.

Question: Why do people feel weightless when they are orbiting?

Lauren: When a spaceship is flying round and round the Earth, it's called orbiting the Earth. The pull of gravity between the spaceship and the Earth keeps the spacecraft from flying away. People in the spaceship will float around weightlessly, even if they are not very far above the Earth.

Session 2.4 Student Reading, CD-ROM Page 4

Imagine shooting a cannon ball out of a cannon on Earth. The cannon ball would shoot out straight, but it would curve its way back to Earth. The faster you shoot out the cannon ball, the farther it would go before falling to Earth. If it shot out super fast, it would go part of the way around the Earth before falling. But if you could shoot the cannon ball out fast enough, the cannon ball would curve around the Earth and fall towards Earth. But it wouldn't land on the Earth! It would keep falling around the Earth as it orbited over and over again.

That's what is happening to astronauts in spacecraft orbiting the Earth. They are falling all the time, without ever landing. They are falling, but they are not falling straight down towards the ground. They're falling mostly sideways, like you've seen in movies where a car goes "flying" off a cliff. But people in space are so high and moving sideways so fast, that they actually "fly" around the whole Earth and never land. That's what orbiting is.

They feel weightless even though the pull of gravity between them and the Earth is still there. If they keep going that fast, they will never fall to Earth. But if they slow down, they will fall to Earth.

Question: Do astronauts feel gravity when they are flying through space?

Lauren: The closer you are to a planet or moon, the more you feel the pull of gravity between you and it. The farther away, the less you feel it. But anywhere you go in space, there is always some gravity, even if it's not very strong. Sometimes astronauts in spaceships can feel like there is no gravity, which means they can actually "float" inside their ship. Sometimes they feel like there is more gravity, which means they feel very heavy. When the spaceship is close to a planet or a moon, or if it is changing its speed, it can feel like there is more gravity. When the space ship is far enough from the planet or moon and is going along at the same speed, it can feel like there is no gravity. When a space ship is moving in the same direction of gravity, like when it's falling, it can also feel like there is no gravity. But whether or not they feel it more or less, they are always being acted on by gravity. Gravity is everywhere!

MEASURING WEIGHTLESSNESS

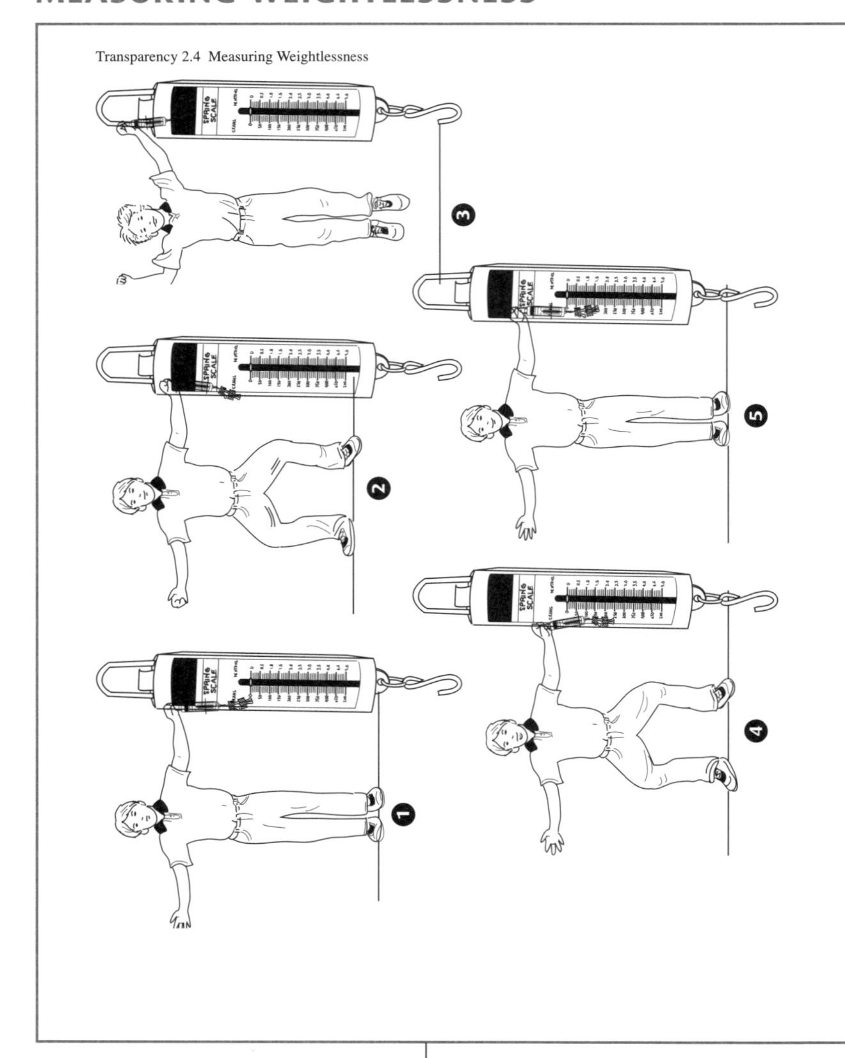

Measuring Weightlessness

1. Hold up a spring scale with two large binder clips attached. Tell your students that they will now measure the effect of movement on weight. Ask a student volunteer to hold the spring scale in front of the class, with the two large binder clips attached. Remind them that a spring scale measures the gravitational pull between the binder clips and the Earth. Get the class to agree approximately where the pointer is pointing on the spring scale.

2. Draw on the overhead transparency where the pointer is pointing. Show the overhead transparency of the *Measuring Weightlessness* sheet. On Figure #1 of the spring scale, record the starting measurement with a transparency pen.

3. Predict what will happen when the student jumps. Ask what they think will happen to the measurement if the student jumps up in the air while holding the spring scale with the binder clips suspended. Take a few predictions.

4. Student jumps while class carefully observes. Tell the student to jump. Repeat this a few times, allowing students to share what they notice.

5. Students focus on observing beginning of jump. On the overhead transparency, direct their attention to Figure #2. Tell them to focus on the very first movement of the pointer on the spring scale. Ask the student to jump again, while the class carefully observes. Repeat this a few times if necessary. Try to come to agreement on what happens to the pointer, and record it on the overhead transparency.

6. The class observes and records five stages of jump. Now, do the same with all five stages of the jump.

Unit Goals

Earth, like all planets and stars, is round like a sphere.

Earth's gravity pulls things toward the center of Earth.

Gravity is everywhere.

TEACHER CONSIDERATIONS

TEACHING NOTES

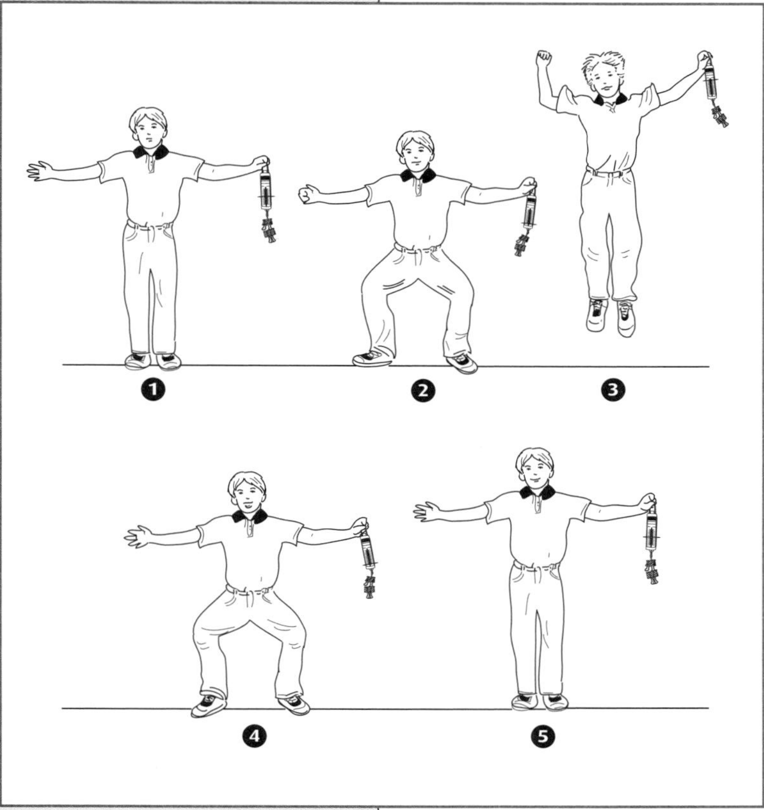

Precise measurements on the spring scale are not necessary: It is not important that all students agree on or even notice exactly where the pointer is in each of the phases of the jump. The point is to observe that the measurement of the weight of the object changes as it is moving up or down (especially falling), even though the mass of the object remains the same.

It is difficult to get accurate numerical readings while the spring scale is in motion, but recording the general position of the pointer or numerical approximations from students who are close enough to read the numbers will work fine.

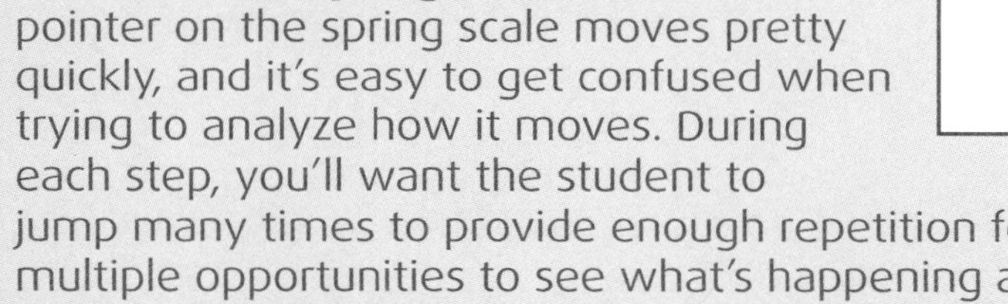

How will the spring scale move? The pointer on the spring scale moves pretty quickly, and it's easy to get confused when trying to analyze how it moves. During each step, you'll want the student to jump many times to provide enough repetition for students to have multiple opportunities to see what's happening and to confirm or dismiss their previous observations. The students will probably notice some or all of the following:

As the student jumps up, the pointer
- starts out at the normal weight (POINTER MIDDLE),
- briefly goes to a weight higher than normal (POINTER DOWN),
- then briefly goes to a weight much lower than normal. (POINTER UP).

When the student lands, the pointer
- briefly goes to a weight higher than normal (POINTER DOWN),
- and returns to the normal weight (POINTER MIDDLE).

***Optional:* Class listens for clicks during jump:** Usually, a spring scale will make a clicking sound when the pointer hits either the top or bottom of the scale. You can ask students to close their eyes and focus on their auditory observations, counting how many clicks they hear during a jump (usually it's three). Then, they can open their eyes and watch to see where the pointer is during each click.

Putting It All Together

1. Invite students to try to explain each motion. Ask students to attempt to explain what caused each motion of the spring scale. What is making it read 0? [It reads 0 when the person and the spring scale are both coming down.] What is making it read the highest weight on the scale? [It reads the highest weight on the scale when the person is jumping up.]

2. There is still gravitational pull between the Earth and the object. Remind them that the gravitational pull between the object and the Earth is not changing. The mass of the object is not changing. It's just the *measurement* that is changing.

3. *Optional:* Show CD-ROM footage of weightlessness in space and in the Vomit Comet. If possible, show footage of astronauts in space and people in the Vomit Comet on the CD-ROM. Use the footage to review the concepts they've learned about gravity.

Unit Goals

Earth, like all planets and stars, is round like a sphere.

Earth's gravity pulls things toward the center of Earth.

Gravity is everywhere.

TEACHER CONSIDERATIONS

QUESTIONNAIRE CONNECTION

The activities in this session deal directly with statements B, C, and D under question #2 on the *Pre–Unit 2 Questionnaire*. At the end of this session, you might want to discuss students' thinking about which of these statements about gravity are true. It is probably best to wait to discuss statements A, E, and F under question #2 until later in the unit.

> 2. Which are true statements? Circle all that are true.
>
> A. The Moon has no gravity because it has no air.
> B. Gravity is an invisible force.
> C. There is no gravity in space.
> D. There is a pull of gravity between all objects.
> E. If there were no air on Earth, people would float out into space.
> F. Gravity keeps the Moon in Earth's orbit.

ASSESSMENT OPPORTUNITY

Critical Juncture: If you notice during the activities and discussions that some students continue to confuse weightlessness with a lack of gravity, you may want to try one or more of the ideas in *Providing More Experience*, which follows.

PROVIDING MORE EXPERIENCE

1. Thought Experiment: Falling while standing on a scale. Tell students to imagine the following:

> You are standing on a scale that is on top of a trap door. The trap door opens, and both you and the scale start falling. Your feet are still on the scale. You look at the scale, and it says, 0. That means that you are weightless. Ask, "Did the gravitational pull between you and the Earth go away?" [No, it is still the same. In fact, that's why you're falling!] "Do you still have the same mass?" [Yes.]

2. Weigh Yourself in an Elevator. This activity is for students who are interested in further investigations outside of school, and who have access to a bathroom scale and an elevator (a non digital scale is needed, as digital scales only take a "snapshot" of your weight and won't allow you to see variation as you would by watching a needle move). Fast elevators in very tall buildings usually show more change.

Continued on page 311

Key Vocabulary

Science and Inquiry Vocabulary

Evidence
Scientific Explanation
Model
Scale Model
Prediction
Scientist
Three–Dimensional (3-D)
Two–Dimensional (2-D)

Space Science Vocabulary

Atmosphere
Air Resistance
Force
Mass
Gravity
Weightless
Satellite
Orbit
Diameter
Sphere
System

Unit Goals

Earth, like all planets and stars, is round like a sphere.

Earth's gravity pulls things toward the center of Earth.

Gravity is everywhere.

PROVIDING MORE EXPERIENCE, *CONTINUED*

a. Before getting into an elevator, set the bathroom scale on a floor and measure and record your weight.
b. Set the scale on the floor of an elevator. Stand on the scale.
c. Watch the scale closely just as the elevator first starts to go up. You will need to have a quick eye. The weight change will not be very big.
d. Do the same thing when the elevator begins going down.
Optional: Without the scale, try jumping in the air at the moment the elevator starts to drop.

Questions

1. What happened to your weight when the elevator began going up?
2. What happened to your weight when the elevator began going down?
3. What do you think would happen to your weight if the elevator broke and you were inside it, dropping fast? (Don't worry: elevators have safeguareds to keep this from happening.)
4. How did it feel when you jumped as the elevator dropped? Was it easier than usual to jump?

Newton's Apple website on weightlessness
http://www.ktca.org/newtons/12/gravity.html

OPTIONAL PROMPTS FOR WRITING OR DISCUSSION

You may want to choose one or more of the prompts below for science journal writing in class or as homework. These prompts could also be used for a discussion or during a final student sharing circle.

- I didn't know this about weightlessness before:_______.
- An experiment I think would be interesting to try in a weightless situation is:_________.
- I wonder if a person would feel weightless in this situation:_____.

Key Vocabulary

Science and Inquiry Vocabulary

Evidence
Scientific Explanation
Model
Scale Model
Prediction
Scientist
Three–Dimensional (3-D)
Two–Dimensional (2-D)

Space Science Vocabulary

Atmosphere
Air Resistance
Force
Mass
Gravity
Weightless
Satellite
Orbit
Diameter
Sphere
System

SESSION 2.5
Gravity and Air

Overview

The teacher drops a binder clip in front of the class, and students are introduced to some ideas about how fast objects fall on Earth. Pairs of students are given materials and challenged to find a way to get one binder clip to fall more slowly than the other. They demonstrate their inventions for the class and discuss how air resistance affects falling objects.

Next, the teacher simultaneously drops a feather and a hammer as a demonstration. Students discuss why one dropped more slowly than the other. They predict what would happen if the same test were performed on the Moon. They watch a brief film clip of an astronaut performing this test on the Moon, and see the feather fall to the ground just as fast as the hammer does.

Students' discussions of the lunar experiment revolve around issues about whether or not there is air or gravity on the Moon. A closing discussion confirms that the Moon does have gravity, but no air. In the next session, the students will gather more evidence about gravity through a simulated Apollo 11 trip to the Moon.

2.5 Gravity and Air	Estimated Time
How objects fall on Earth	5 minutes
Investigations with gravity	15 minutes
Discussing the gravity investigations	10 minutes
A falling hammer and feather on Earth	10 minutes
The hammer and feather experiment on the Moon	10 minutes
Discussing the Moon experiment and introducing key concepts about gravity and air	10 minutes
TOTAL	**60 minutes**

What You Need

For the class

- ❑ sentence strips for three key concepts
- ❑ wide-tip felt pen
- ❑ 1 feather (A fluffy feather is best, rather than a long, rigid quill. It is best if the feather is no longer than 10 inches.)
- ❑ 1 hammer, any kind
- ❑ 1 newspaper
- ❑ 1 grain of sand or tiny pebble
- ❑ 1 piece of white paper or cloth, at least 2 feet square

Unit Goals

Earth, like all planets and stars, is round like a sphere.

Earth's gravity pulls things toward the center of Earth.

Gravity is everywhere.

What Some Teachers Said

"The kids LOVED this lesson. They had so much fun making the devices. Many of them used the plastic bags to make parachutes. One team simply laid the binder clip on top of a folded paper towel. It still worked."

"This session was fun for the students. They loved making the contraptions to slow the binder clip."

Key Vocabulary

Science and Inquiry Vocabulary

Evidence
Scientific Explanation
Model
Scale Model
Prediction
Scientist
Three–Dimensional (3-D)
Two–Dimensional (2-D)

Space Science Vocabulary

Atmosphere
Air Resistance
Force
Mass
Gravity
Weightless
Satellite
Orbit
Diameter
Sphere
System

- ❑ 16 lightweight plastic shopping bags with handles (plastic grocery bags are ideal)
- ❑ about 16 sheets of scratch paper
- ❑ a few paper bags, any size
- ❑ a roll of masking tape
- ❑ a few yards of string
- ❑ several pairs of scissors
- ❑ a copy of the Apollo 15 feather and hammer drop footage, either on CD-ROM or VHS tape
- ❑ large screen monitor/LCD projector or VCR to show feather/hammer video clip

For each pair of students

- ❑ 2 small binder clips, 1 ¼ inch wide or smaller

Getting Ready

1. Gather the materials students can use in slowing the fall of a binder clip, and set them in a central location. It is helpful if materials are spread out over a long counter or table for easier access. Plan to have students share materials such as scissors, tape, and string:

- 16 lightweight plastic shopping bags with handles
- at least 16 sheets of paper
- about 10 paper bags
- tape
- string
- scissors

2. Have enough small binder clips handy so that each pair of students will get two.

3. Set up the LCD projector, screen, or VCR to show the feather and hammer film clip. (Please see note on right-hand page, 315.)

4. Set out the piece of white paper or cloth and a grain of sand or very small pebble. Also have handy the hammer and feather, and a pad of newspaper on which you can drop them.

5. Use the wide-tip felt pen to write the following key concepts on

Unit Goals

Earth, like all planets and stars, is round like a sphere.

Earth's gravity pulls things toward the center of Earth.

Gravity is everywhere.

CD–ROM NOTES

Feather—Hammer Video on the Moon

The whole class needs to see the video: During this session, you will show a short (about 1 minute) NASA video of an astronaut dropping a hammer and a feather on the Moon during the Apollo 15 mission in 1971. The movie is provided on the CD-ROM. Because this short movie is an integral part of the learning in this session, teachers who do not have the equipment necessary to show the CD-ROM to the whole class will need to obtain a VHS version (supplied in materials kit). Alternatively, you could rotate pairs of students to a computer to view the movie throughout several days.

8.3 Mb Quicktime movie of the demonstration
http://nssdc.gsfc.nasa.gov/planetary/lunar/apollo_15_feather_drop.html

If you are using the CD-ROM version: Preview it before class. Instructions for using this program are included on the CD–ROM.

The quality of this historic and fascinating movie is somewhat grainy. Preview it before class and plan how to dim the lights, and so on, for optimal viewing. Also, plan to show it more than once, so that students are sure to see the feather and hammer hit the ground at the same time.

Key Vocabulary

Science and Inquiry Vocabulary

Evidence
Scientific Explanation
Model
Scale Model
Prediction
Scientist
Three–Dimensional (3-D)
Two–Dimensional (2-D)

Space Science Vocabulary

Atmosphere
Air Resistance
Force
Mass
Gravity
Weightless
Satellite
Orbit
Diameter
Sphere
System

sentence strips, and have them ready to post on the concept wall at the end of the session:

Air resistance acts against the movement of objects through air.
There is no air on the Moon or in space.
There is gravity on the Moon.

How Objects Fall on Earth

1. Drop a binder clip. As the class watches, drop a binder clip from about the height of your head. Ask what made it fall to the ground. [Gravity.] Ask, "What would happen if someone in Australia dropped the clip?" [It would fall to the ground.] "Why?" [The gravitational pull between Earth and other objects pulls objects toward the center of the Earth.]

2. Watch how fast it drops. Tell students that you're going to drop it again, but this time they should focus on how fast the clip falls.

3. Objects fall at this speed when dropped from this height. Drop the clip again and point out that it drops at the same speed every time. Tell students that this is how fast objects drop on Earth when dropped from this height.

4. Gravity is the same everywhere on Earth. Tell students that no matter where the object is dropped on the planet, from this same height it would always fall at this same speed, because the gravitational pull between the Earth and other objects is the same all over the planet.

5. Objects speed up as they fall farther. Tell them that the farther something falls, the faster it falls. The same binder clip would speed up more if dropped from a higher point.

Unit Goals

Earth, like all planets and stars, is round like a sphere.

Earth's gravity pulls things toward the center of Earth.

Gravity is everywhere.

TEACHER CONSIDERATIONS

SCIENCE NOTES

Air resistance: It isn't strictly accurate to say that air resistance "slows an object's fall." Scientists would say that air resistance slows the object's *acceleration*. When an object is dropped, air resistance acts against the acceleration due to the pull of gravity. The object that was dropped will still gain speed as it falls, but it won't speed up as quickly as it would if there were no air present. Air resistance works directly against the movement of an object, in whatever direction the object is moving, and not only when it is falling. For upper elementary students, it is not necessary to explain this fully, but, to be accurate, we have phrased the key concept, "Air resistance acts against the movement of objects through air." In the activity, however, we use the phrase "slow an object's fall" because this phrasing seems most pertinent to the situations that students are considering.

Key Vocabulary

Science and Inquiry Vocabulary

Evidence

Scientific Explanation

Model

Scale Model

Prediction

Scientist

Three–Dimensional (3-D)

Two–Dimensional (2-D)

Space Science Vocabulary

Atmosphere

Air Resistance

Force

Mass

Gravity

Weightless

Satellite

Orbit

Diameter

Sphere

System

Investigations of Gravity

1. **Challenge students to slow the fall of a binder clip.** Say that they will work in teams of two to see if they can figure out ways to fight gravity—to slow a binder clip down as it falls.

2. **Each pair of students will get two binder clips.** They will drop one clip just as it is, but they will get to attach some materials to their other clip to see if they can make it fall more slowly. This way, they can compare the speed at which the two clips fall.

3. **Show them the available materials.** Point out the plastic bags, paper bags, paper, string, tape, and scissors. Explain that they will decide on one or two materials to use to try to slow down their clip as it falls. Tell them that they will only have a few minutes, so they will probably have time to try only one way to slow the binder clip's fall.

4. **Divide into pairs.** Say that pairs of students will work together. They should talk over what materials they want to use, and then one partner should get the materials. Give the class an idea of how long they have for the activity. Caution them that the objective isn't to make a fancy device, but to try out an idea.

5. **Go!** Give each team two binder clips, and allow them to plan, gather materials, and go to work.

Discussing the Gravity Investigations

1. **Regain their attention.** When it seems that most of your students have come up with a strategy for slowing a binder clip's fall, collect the materials and have students clean up. If students have particularly good devices they want to share, let them keep them for the discussion.

2. **Students share their strategies.** Ask students to share the strategies they devised. Have a few volunteers bring their devices to the front of the room to share. Remind them to drop the device with one hand as they simultaneously drop a binder clip without the device (the "control") from the same height with the other hand.

3. **Air resistance is slowing their fall.** Note that their devices are indeed making the binder clips fall more slowly. Ask what it is that is making them fall more slowly. If no one mentions it, introduce the term *air resistance.*

Unit Goals

Earth, like all planets and stars, is round like a sphere.

Earth's gravity pulls things toward the center of Earth.

Gravity is everywhere.

TEACHER CONSIDERATIONS

TEACHING NOTES

What to expect: Students enjoy the investigations with gravity and air resistance and are usually successful. Although they may come up with many creative strategies for slowing the binder clip, it is likely that one of the most successful will be a kind of parachute. Provide only 10 to 15 minutes for this activity, and tell students that they need to devise only one way of slowing the binder clip. Although they would probably enjoy spending much more time devising a variety of contraptions, a short experience is all they need for the discussion that follows.

Key Vocabulary

Science and Inquiry Vocabulary

Evidence

Scientific Explanation

Model

Scale Model

Prediction

Scientist

Three–Dimensional (3-D)

Two–Dimensional (2-D)

Space Science Vocabulary

Atmosphere

Air Resistance

Force

Mass

Gravity

Weightless

Satellite

Orbit

Diameter

Sphere

System

The Falling Feather and Hammer

1. Predict whether a feather and a hammer will fall at the same speed. Show your students the feather and the hammer. Tell them that you're going to drop them at the same time from the same height. Ask for their predictions of whether they will fall at the same speed, or if one will fall faster than the other.

2. Drop the feather and hammer. Choose a place on the floor or a table that everyone can see. Put a pad of newspaper or other padding on the surface where the hammer will fall. Make sure that you are holding the entire hammer and the entire feather at approximately the same height, and then drop them at the same moment. [The feather falls more slowly.]

Note: If you have a long feather and drop it pointed end first, the streamlined shape may not provide much air resistance. Try dropping it side first.

3. Air resistance slows the feather. Repeat the demonstration a few times and then ask, "Why did the feather fall more slowly?" [Air resistance slowed its fall.] Point out that this is similar to when they slowed the fall of the binder clip by increasing air resistance.

4. Many people think that the feather falls more slowly because it has less mass. It is likely that some students will think that the feather falls more slowly because it is lighter than the hammer. In other words, they may think that if an object has more mass, it will fall faster than an object with less mass. Tell them that that many people think that this is the case. Say that this idea can be tested.

5. Predict whether a sand grain or a feather will fall faster. Hold a grain of sand or a tiny pebble in one hand and the feather in the other. Point out that the grain of sand has less mass and weighs less than the feather. Tell them you are going to drop them at the same time, and see which lands first. Ask for predictions.

6. Drop the sand grain and feather over the white paper and observe. Hold the two objects above a piece of white paper or cloth, so the grain of sand can be seen when it lands. Make sure that you are holding the grain of sand and the entire feather at approximately the same height, and then drop them at the same moment. [The feather falls more slowly.]

Unit Goals

Earth, like all planets and stars, is round like a sphere.

Earth's gravity pulls things toward the center of Earth.

Gravity is everywhere.

TEACHING NOTES

Hold off on discussion about how objects on Earth would fall without air resistance: A student may point out that if there were no air, all objects on Earth would fall at the same rate. If this comes up, confirm that this is the case, although it surprises most people. If no one brings it up, wait to discuss this issue until after students have observed the footage of the hammer and feather drop on the Moon. The footage is a very persuasive way to illustrate the idea that objects of different mass fall at the same rate when there is no air resistance.

What One Teacher Said

Despite discussions we've had about gravity and air, most students thought that the hammer and feather would fall slower than on the Earth because the Moon has less gravity, but that the feather would still fall more slowly than the hammer. They were very surprised to see both the hammer and the feather fall at the same speed!

Key Vocabulary

Science and Inquiry Vocabulary

Evidence
Scientific Explanation
Model
Scale Model
Prediction
Scientist
Three–Dimensional (3-D)
Two–Dimensional (2-D)

Space Science Vocabulary

Atmosphere
Air Resistance
Force
Mass
Gravity
Weightless
Satellite
Orbit
Diameter
Sphere
System

7. Repeat the demonstration. Air resistance, not less mass, slows feather. Because the grain of sand is hard to see, you will need to repeat the demonstration a few times. Ask, "Which object has more mass (weighs more)?" [The feather.] "Which object fell faster?" [The grain of sand.] "Why did the feather fall more slowly?" [Air resistance.]

8. Why did the hammer fall faster than the feather? The feather meets with more air resistance. Hold up the hammer and feather again, and drop them simultaneously again. *Tell them that the reason the hammer falls faster than the feather is not because it has more mass; It is because the feather meets with more air resistance.* (See note on right-hand page, 323.)

What Would Happen on the Moon?

1. What would happen if the feather and hammer were dropped on the Moon? Tell students that this test was actually performed on the Moon by an astronaut, and they will get to see the film footage in a few minutes. Ask students to discuss with a neighbor what they think would happen if a feather and a hammer were dropped by a person standing on the Moon.

2. Circulate and listen to their discussions. Encourage them to apply everything they have learned about gravity so far as they discuss this question. Note whether students think that there is gravity on the Moon, and whether they think that there is air. Don't tell them yet if there is air or gravity on the Moon.

3. Show the film clip. Show the footage of the astronaut dropping the feather and the hammer. (The two objects fall to the ground at exactly the same rate!) You may choose to repeat the clip a few times.

Discussing Why the Feather and Hammer Fell at the Same Rate on the Moon

1. Discuss what they saw. Ask if they were surprised by the results, and if so, what surprised them. Ask, "Why did the feather fall at the same rate as the hammer when on Earth the feather drops more slowly?" "What conditions might be different on the Moon to cause this?" "Is there gravity on the Moon?" [Yes, the objects fell to the ground.] "Is there air on the Moon?" [No, because the objects fell at the same rate: there was no evidence of air resistance on the feather.]

You may want to mention that the pull of gravity on the Moon is less than on Earth, because the Moon has less mass than the Earth. However, this will also come up in Session 2.6.

Unit Goals

Earth, like all planets and stars, is round like a sphere.

Earth's gravity pulls things toward the center of Earth.

Gravity is everywhere.

TEACHER CONSIDERATIONS

TEACHING NOTES

Convincing students that objects would fall at the same rate on Earth: Students may challenge the idea that, barring the effects of air resistance, all objects fall at the same rate on Earth. It is probably best to tell them this is what scientists have found, but you may want to leave this concept for deeper exploration in higher grades. Without careful instruments and monitoring, this is difficult to demonstrate in class. A young student's bias toward heavier objects falling faster and hitting the ground first is often so strong that even when they witness two objects hitting the ground at the same time, they often interpret it as the heavier object having hit first.

If you decide to pursue further the idea of objects falling at the same rate, you may want to do more tests dropping objects of different mass. If you have a multistory school and a stopwatch, try dropping the binder clip from 10 meters high while timing it. However, be sure to pretest the objects you choose. For example, two similar-sized balls of different weights might work well. If you drop an object with less mass but with slightly more air resistance than the other object, some students may mistakenly conclude that objects with more mass do fall more quickly than objects with less mass.

QUESTIONNAIRE CONNECTION

4. If a person is standing on the Moon holding a rock and then lets go of it, what will happen to the rock? It will __________

A. Float away.
B. Float where it is.
C. Fall to the ground.
D. Go up.

Question #4 on the *Pre–Unit 2 Questionnaire* focuses on what would happen to a rock when dropped on the Moon. After watching and discussing the hammer and feather drop on the Moon, students should be convinced that answer C (the rock will fall to the ground) is the correct response to Question #4. However, if you notice any confusion during the discussion, take some time to review the question.

What Some Teachers Said

"After our discussion from lesson 3 about gravity and other planets etc., my students were amazed about the feather and hammer demonstration. They loved the idea that it was live footage! Their ideas about air and gravity and space and gravity are very clear now."

"This was a great session. The students really enjoyed the discussions and the hands-on activity. Seeing the hammer/feather experiment done before their eyes and then seeing the same experiment done on the Moon was a great way for them to grasp how the Moon and Earth environments affect things differently."

2. A rock would also drop on the Moon. Reveal that if a rock were dropped on the Moon, it would also fall in the same way and at the same speed. So would a binder clip. On the Moon, the contraption they designed to slow down a binder clip with air resistance would drop in the same way and at the same speed as a binder clip alone.

3. Without air resistance, objects would fall at the same rate on Earth. Ask, “If there were no air on Earth, would the sand grain and the feather fall at the same rate?” [Yes, because there would be no air resistance.]

Post Key Concepts About Gravity and Air

1. Post the three key concepts below on the concept wall and read them aloud. Tell the class that they will learn more about gravity and air in Ssession 2.6, which is about the Apollo 11 mission, during which people set foot on the Moon for the first time.

Air resistance acts against the movement of objects through air.
There is no air on the Moon or in space.
There is gravity on the Moon.

Unit Goals

Earth, like all planets and stars, is round like a sphere.

Earth’s gravity pulls things toward the center of Earth.

Gravity is everywhere.

TEACHER CONSIDERATIONS

PROVIDING MORE EXPERIENCE

About air resistance: The reading about skydiver Joseph Kittinger, *Jumping from the Edge of Space* (Unit 1, Session 1.8, from the student sheet packet), provides an interesting personal account of falling toward Earth from high in the atmosphere (where there is little air), passing through lower altitudes (where there is more air), and finally, using a parachute to slow his fall. If your students did not read this story in Unit 1, you could have them read it now, or you could just review some key points from the story having to do with gravity and air resistance:

- In 1960, a man named Joseph Kittinger jumped from a balloon 31,000 kilometers high.
- Joseph fell very quietly at first, because there was so little air.
- Because there was very little air resistance up so high, Joseph fell much faster than skydivers, who jump from lower altitudes.
- When Joseph got low enough in the atmosphere, where there is more air, he used his parachutes.

OPTIONAL PROMPTS FOR WRITING OR DISCUSSION

Choose one or more of the prompts below for science journal writing in class or as homework. They could also be used for a discussion or during a final student sharing circle.

- The best device we made for slowing down the binder clip was:___________.
- I think the reason it slowed the binder clip down was:_________.
- An experiment I think would be interesting to try on the Moon is:_______________. I predict that this would happen:___________.
- I think that it would be interesting to drop these things on the Moon:___________. These are my predictions of what would happen to each one:_______________.
- These are things I wonder about gravity and air:__________.

What One Teacher Said

"The CD–ROM was a hit! The students were especially engaged during this part of the lesson and had so many questions. It was fantastic! So many had the impression before this lesson that we would just float away on the moon. BEST lesson so far!"

Key Vocabulary

Science and Inquiry Vocabulary

Evidence
Scientific Explanation
Model
Scale Model
Prediction
Scientist
Three–Dimensional (3-D)
Two–Dimensional (2-D)

Space Science Vocabulary

Atmosphere
Air Resistance
Force
Mass
Gravity
Weightless
Satellite
Orbit
Diameter
Sphere
System

SESSION 2.6
Gravity Beyond Earth

Unit Goals

Earth, like all planets and stars, is round like a sphere.

Earth's gravity pulls things toward the center of Earth.

Gravity is everywhere.

Overview

The final session in Unit 2 begins with a simulated mission to the Moon, as the students are shown a series of 21 images from the Apollo 11 mission. After viewing the mission, students review the concept of gravity by discussing two questions in evidence circles, and backing up their answers with evidence from the Apollo 11 images: Is there gravity on the Moon? Is there air on the Moon? The evidence that there is indeed gravity on the Moon, but no air, is shared in a class discussion. Students deconstruct the common misconception that air is necessary for gravity to exist. The Apollo 11 images also serve to reinforce the concept of the spherical shape of the Earth and Moon.

To conclude Unit 2, students take the *Post–Unit 2 Questionnaire* to find out how their ideas have changed since the beginning of the unit.

2.6 Gravity Beyond Earth	Estimated Time
Apollo 11 Mission to the Moon	20 minutes
Evidence circles on air and gravity	15 minutes
Discussion about air and gravity	10 minutes
Taking the *Post–Unit 2 Questionnaire*	15 minutes
TOTAL	**60 minutes**

What You Need

For the class

- ❑ overhead projector or computer with large screen monitor/LCD projector
- ❑ transparencies from the transparency packet or CD-ROM file of the *Apollo 11 Mission to the Moon* images
- ❑ wide-tip felt pen
- ❑ sentence strip for 1 key concept

For each student

- ❑ *Post–Unit 2 Questionnaire* from the student sheet packet
- ❑ *optional:* lined paper for optional writing assignment about air and gravity on the Moon

Getting Ready

1. Arrange for the appropriate projector format (computer with large screen monitor, LCD projector, or overhead projector) to display images to the class.

THE KEY CONCEPT WALL

BY THE END OF UNIT 2, YOUR CONCEPT WALL SHOULD LOOK SOMETHING LIKE THIS:

WHAT WE HAVE LEARNED ABOUT EVIDENCE AND MODELS

1. Evidence is information, such as measurements or observations, that is used to help explain things.
2. Scientists base their explanations on evidence.
3. Scientists question, discuss, and check each other's evidence and explanations.
4. Scientists use models to help understand and explain how things work.
5. Space scientists use models to study things that are very big or far away.
6. Models help us make and test predictions.

Note: These three concepts are not introduced in Units 2, 3, and 4. However, if you introduced them in Unit 1, keep them on the concept wall:

7. Every model is inaccurate in some way.
8. Models can be 3-dimensional or 2-dimensional.
9. A model can be an explanation in your mind.

WHAT WE HAVE LEARNED ABOUT SPACE SCIENCE

2.2

The Earth, like all planets and stars, is round, like a sphere.

The Earth is so big that we can't see that it's round when we're on it.

People live all over the surface of the Earth.

2.3

Gravity is an invisible pulling force.

All objects have gravity that pulls on all other objects.

Objects with more mass have more gravity.

The farther apart objects are, the less the pull of gravity between them.

Earth's gravity pulls things toward the center of the Earth.

It is hard to leave Earth because of the pull of Earth's gravity.

Gravity can pull across great distances.

People have been gathering evidence about space for a long time.

Current missions continue to provide evidence about space.

2.4

Gravity is everywhere.

People feel weightless in some situations but there is still a pull of gravity between them and other objects.

2.5

Air resistance acts against the movement of objects through air.

There is no air on the Moon or in space.

There is gravity on the Moon.

2.6

Air is not necessary for gravity to exist.

2. If you will not be using the CD-ROM, make overhead transparencies of the 21 *Apollo 11 Mission to the Moon* images.

3. Make one copy of the *Post–Unit 2 Questionnaire* for each student.

4. *Optional:* Provide lined paper if you will have students record the evidence that there is gravity on the Moon, but no air (see *Optional Embedded Assessment*, page 343).

5. Write the following key concept on a sentence strip, and have it ready to post on the concept wall during the session.

Air is not necessary for gravity to exist.

Apollo 11 Mission to the Moon

1. Images from Apollo 11. Tell the class that they'll be seeing a series of 21 images from Apollo 11, the first mission that took people to the Moon. Tell them that the images provide evidence that may help to answer their questions about gravity.

2. Evidence circles will discuss some of the images. Divide the class into evidence circles of approximately four students each. Let students know that they will get a chance to use some of the images as evidence to answer questions about air and gravity on the Moon.

3. Show the *Mission to the Moon* images. Show the series of images. Using the teacher script, describe briefly what students are seeing in each photo, thus keeping the "mission" moving along and preventing interest from waning.

Apollo 11 Mission to the Moon Teacher Script

Image #1: Astronaut inside capsule

In 1969, people had made trips into space, but no person had yet been to the Moon. This is one of the astronauts inside the space capsule. Note how small the space capsule is. The capsule was designed to be as small as possible, because the more mass something has, the harder it is to lift it enough to escape the gravitational pull between it and the Earth.

More information: This is a photo of Command Module Pilot Michael Collins in the command module simulator during a simulated rendezvous and docking maneuver.

Unit Goals

Earth, like all planets and stars, is round like a sphere.

Earth's gravity pulls things toward the center of Earth.

Gravity is everywhere.

TEACHER CONSIDERATIONS

CD-ROM NOTES

CD-ROM version of the Apollo 11 Mission to the Moon: If you are using the CD-ROM version, preview it before class. Click the "back" and "next" buttons with your mouse to journey chronologically through this series of photos from the Apollo 11 mission. Look at astronauts training, liftoff, the landing on the Moon, splashdown, and the astronauts back on Earth. Further instructions for using this program are included on the CD–ROM.

Apollo Videos—Optional resource on the CD-ROM for independent student exploration: On the CD–ROM, you will also find several short movies. These are not designed to be used with the whole class, but rather as a student-centered enrichment opportunity if you have a computer available for students to use individually or in pairs. View video clips of astronauts exploring, at work, or at play, by clicking on the numbers in the lower left-hand corner of the screen. Informational text about the Apollo missions and the Moon appears to the right of the video. To start another video, just click another numbered button with your mouse. All videos were taken from the Apollo 15, 16, and 17 missions. Further instructions for using this program are included on the CD–ROM.

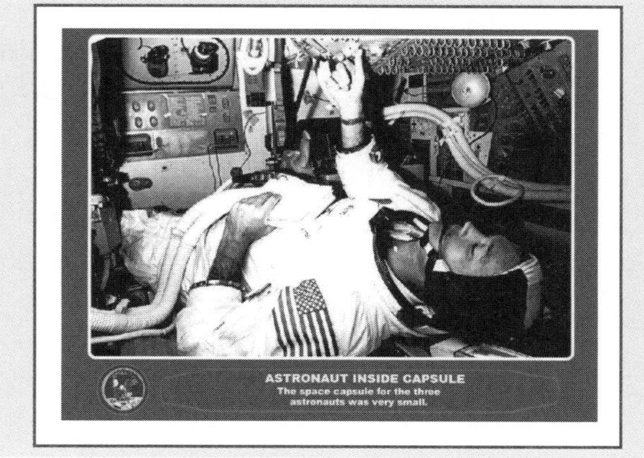

Key Vocabulary

Science and Inquiry Vocabulary

Evidence
Scientific Explanation
Model
Scale Model
Prediction
Scientist
Three–Dimensional (3-D)
Two–Dimensional (2-D)

Space Science Vocabulary

Atmosphere
Air Resistance
Force
Mass
Gravity
Weightless
Satellite
Orbit
Diameter
Sphere
System

Image #2: Liftoff

The rocket was big because it takes a very powerful rocket with big fuel tanks to escape the gravitational pull between it and the Earth.

More information: This is the rocket as the mission launched on July 16, 1969. It was a clear, sunny Wednesday at Kennedy Space Center, Florida.

Images #3 & #4: Liftoff

The astronauts were inside the cone-shaped part at the front of the rocket, below the thin, pointy part.

More information: The second photo was taken by a camera on the launch tower. Most of the huge rocket was used for the power to push them away from the pull of gravity between the rocket and Earth; the rocket later broke off.

Image #5: The whole Earth

When they were far enough away, in orbit, the astronauts were able to see the whole Earth. *What shape is it?* This far from Earth, the astronauts felt weightless and could float around the spacecraft. Weightlessness does not mean that there is no gravity. (If appropriate, explain that orbit is really a free-fall around the Earth. Even though there is gravity, they felt weightless because of their orbital motion, as described in Session 2.3 and in *Background for Teachers.*)

More information: The Apollo spacecraft reached Earth parking orbit after 11 minutes. After one and a half orbits, the thrusters fired and the astronauts headed toward the Moon. This photo was taken from 158,000 km (98,000 miles) away from Earth. The Moon is 384,400 km (240,000 miles) from Earth.

Image #6: The whole Moon

The Moon looked bigger and bigger as the astronauts got closer and closer to it. *Was the Moon really getting bigger, or did it just look bigger? What shape is it?*

More information: This photo was actually taken on their return flight.

Unit Goals

Earth, like all planets and stars, is round like a sphere.

Earth's gravity pulls things toward the center of Earth.

Gravity is everywhere.

LIFTOFF
The rocket was very big. Here the 111 meter tall Apollo 11 spacecraft launches for the Moon.

LIFTOFF
The rocket is lifting off the ground against the pulling force of the gravitational pull between it and the Earth.

THE WHOLE EARTH
Apollo 11 astronauts took this photo of the Earth when the spacecaft was 180,000 km away.

THE WHOLE MOON
Apollo 11 astronauts took this photo of the Moon when the spacecaft was 18,000 km away.

Key Vocabulary

Science and Inquiry Vocabulary

Evidence
Scientific Explanation
Model
Scale Model
Prediction
Scientist
Three–Dimensional (3-D)
Two–Dimensional (2-D)

Space Science Vocabulary

Atmosphere
Air Resistance
Force
Mass
Gravity
Weightless
Satellite
Orbit
Diameter
Sphere
System

Image #7: Earthrise

It took four days to get to the Moon and begin orbiting the Moon. *What do you think you're seeing in this photo?* [This is the Earth seen from the Moon, above the Moon's horizon.]

Image #8: Landing on the Moon

Two of the three astronauts got into the lunar module, which separated from the main spacecraft to land on the Moon. The third astronaut stayed in orbit in the command module. As the lunar module slowed down before landing, the astronauts stopped floating in their spacecraft and sank into their seats. When they landed, this is what they saw out the window.

Pause and ask this important review question: *Why does the Moon look flat in this photo if we know that it's really round like a ball?* [Although the Moon is smaller than the Earth, it is still so big compared with a person that when you're on it, you can't tell that it is round.]

They did not use a parachute to slow their fall to Earth. Instead, they used rockets. *Why didn't they use a parachute?*

More information: "The Eagle has landed," were the first words heard from the two astronauts after they landed in the lunar module (LM). They had to fly longer in the LM than planned to avoid a field of boulders and a crater. They had only 40 seconds of fuel left when they landed. They landed on a part of the Moon called the Sea of Tranquility.

Image #9: The first step

This is a photo of the first person to ever step on the Moon. As he took the step, he said, "That's one small step for man, one giant leap for mankind." He was being watched on TV by more than a half–billion people on Earth.

What he said made no sound on the Moon except inside the helmets and spacecraft, where there is air. This is because sound waves cannot travel without air.

More information: Neil Armstrong took the first step. The photo was taken by a camera mounted on one of the legs of the lunar module.

Unit Goals

Earth, like all planets and stars, is round like a sphere.

Earth's gravity pulls things toward the center of Earth.

Gravity is everywhere.

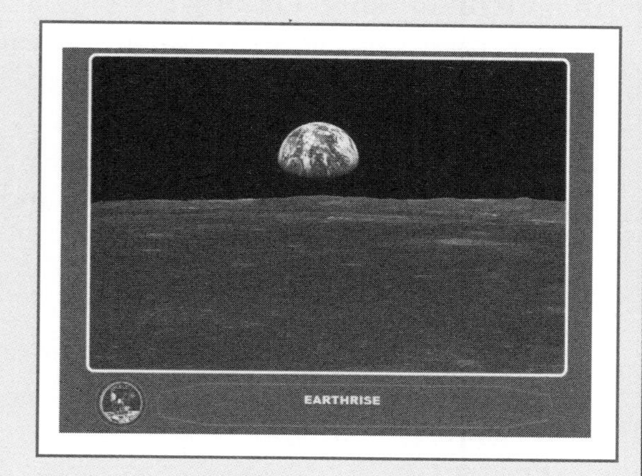

LANDING ON THE MOON
The Apollo 11 landing site on the Moon.

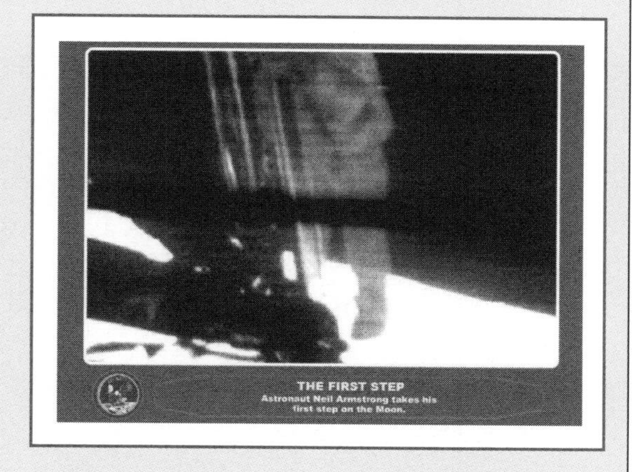

What Some Teachers Said

"Great lesson and photos!"

"My students thought the Apollo footage was fantastic! It really solidified their thoughts on gravity."

"My students loved the CD–ROM pictures. This group of students have not been exposed to anything like this before and were absolutely fascinated. I still had a few students wondering about the air issue on the moon at the end, but overall they 'got it'."

Image #10: Making a bootprint

This photo shows an astronaut making a bootprint in the dust on the Moon.

More information: This was actually part of a planned experiment to study the nature of the lunar dust and the effects of pressure on the surface.

Image #11: A bootprint

Note: For now, no comment is needed. This image will serve as evidence later during a discussion of whether there is air on the Moon.

More information: Some people think that this is a photo of the first bootprint on the Moon, but it isn't.

Image #12: Astronaut on the Moon

Here is a photo of one of the astronauts standing on the Moon in his space suit.
Do you see any evidence in this photo for gravity on the Moon?

More information: This is a photo taken by Neil Armstrong of Edwin "Buzz" Aldrin. Aldrin was busy setting up experiments and didn't think to take a photo of Armstrong, but Armstrong's reflection can be seen on the visor.

Image #13: Astronaut with flag

The flag in this picture looks as though it is waving in the wind. It had a bar across the top to hold it up. Without the bar, it would have hung down on the pole limply.

Unit Goals

Earth, like all planets and stars, is round like a sphere.

Earth's gravity pulls things toward the center of Earth.

Gravity is everywhere.

MAKING A BOOTPRINT
Apollo 11 astronaut makes a bootprint on the Moon.

A BOOTPRINT
Their bootprints will still pobaly look the same even millions of years later.

ASTRONAUT ON THE MOON
Apollo 11 astronaut Buzz Aldrin in his spacesuit on the Moon.

ASTRONAUT WITH A FLAG
Buzz Aldrin with the United States flag. on the Moon.

Key Vocabulary

Science and Inquiry Vocabulary

Evidence
Scientific Explanation
Model
Scale Model
Prediction
Scientist
Three–Dimensional (3-D)
Two–Dimensional (2-D)

Space Science Vocabulary

Atmosphere
Air Resistance
Force
Mass
Gravity
Weightless
Satellite
Orbit
Diameter
Sphere
System

Image #14: Astronaut carrying things

The astronauts were able to carry things on the Moon much more easily than on Earth. *Why do you think this is?*

More information: This is Aldrin carrying experiments to set up.

Image #15: Astronaut jumping

This astronaut, together with his suit and equipment, weighed almost 400 pounds on Earth, but he was able to easily jump this high. *Why do you think it was easier to jump on the Moon than on Earth?* [The astronaut weighed less on the Moon than on Earth (only 65 pounds on the Moon!), so his muscles were able to lift him higher.] Emphasize that he came back down to the ground.

More information: This is actually a photo taken on the Apollo 16 mission. It is John Young jumping as he salutes the flag. The photo was taken by another astronaut facing him.

Image #16: Back in orbit around the Moon

Then Buzz and Neil got back in the lunar module and took off to join the command module, orbiting the Moon. In orbit, they felt weightless again and could float around in their spacecraft. *Can you tell what the three things in this photo are? Does the Moon look round or flat in this picture?*

More information: This is a photo taken from the lunar module as it approached the command module for docking. The Earth is seen in the background above the Moon.

Image #17: Crescent Earth

As the astronauts returned home, Earth looked bigger and bigger. In this photo, it looks like a crescent shape. *Why do you think that the Earth's shape looks like this in this photo? Why are some parts dark and some parts light?* [The Earth appears crescent-shaped because the Sun is shining only from the right side of the photo. The light part of the Earth has daytime, and the other side has night.]

Unit Goals

Earth, like all planets and stars, is round like a sphere.

Earth's gravity pulls things toward the center of Earth.

Gravity is everywhere.

What One Teacher Said

"The students loved the photos of real astronauts. There was extremely high interest in this lesson."

Key Vocabulary

Science and Inquiry Vocabulary

Evidence
Scientific Explanation
Model
Scale Model
Prediction
Scientist
Three–Dimensional (3-D)
Two–Dimensional (2-D)

Space Science Vocabulary

Atmosphere
Air Resistance
Force
Mass
Gravity
Weightless
Satellite
Orbit
Diameter
Sphere
System

Image #18: Closer Earth

As they flew closer to Earth, it looked even bigger. And the closer they got, the more strongly gravity pulled them toward it.

Image #19: Parachuting spacecraft

Once inside Earth's atmosphere, the astronauts could use parachutes to slow down their fall to Earth. *Why won't parachutes work in space? Why do parachutes work in Earth's atmosphere? Why did they need parachutes?*

More information: This is actually a photo of Apollo 16.

Image #20: Floating spacecraft

They landed on the ocean, and were picked up by people in helicopters and on rafts. *Why do you think they landed on water instead of land?* [Softer impact.]

More information: They landed at 12:50 P.M. EDT July 24, 1969, about 812 nautical miles southwest of Hawaii. They wore biological isolation garments and were transported by helicopter to the USS *Hornet*.

Image #21: Happy astronauts

The three astronauts landed safely.

More information:
The astronauts were kept in quarantine for three weeks.

Unit Goals

Earth, like all planets and stars, is round like a sphere.

Earth's gravity pulls things toward the center of Earth.

Gravity is everywhere.

TEACHER CONSIDERATIONS

PROVIDING MORE EXPERIENCE

For more images of Apollo missions direct your students to websites such as these:

http://nssdc.gsfc.nasa.gov/planetary/lunar/apollo11.html

http://www.hq.nasa.gov/office/pao/History/ap11ann/kippsphotos/apollo.html

http://www.hq.nasa.gov/alsj/a11/images11.html

http://images.jsc.nasa.gov/luceneweb/caption.jsp

Apollo image gallery:
http://www.apolloarchive.com/apollo_gallery.html

Other Moon Missions: Your students may become curious about other Moon missions or space missions generally. See *Resources* on pages 57-59 for more information about missions.

Space Bounce: Players move a trampoline to bounce astronauts and their cargo to a lunar base. Provides experience with seeing the speed at which objects would drop on the Moon, and how they would bounce.
http://www.nasa.gov/audience/forkids/games/G_Spacebounce.html

Apollo Lander Game: A game in which players attempt to land an Apollo lunar module on a specific target by firing thrusters.
http://nasaexplores.com/extras/apollo11/moonlander/moonlander.html

Mars Rover Landing Images: If your students found the Apollo mission intriguing, and if they enjoyed designing air resistance contraptions, you may choose to show them a fascinating series of images depicting the stages in the landing of the 2005–2006 Mars Rover missions, which include parachutes and bouncing. These can be found on the Internet at
http://marsrovers.jpl.nasa.gov/mission/tl_entry1.html

Key Vocabulary

Science and Inquiry Vocabulary

Evidence
Scientific Explanation
Model
Scale Model
Prediction
Scientist
Three–Dimensional (3-D)
Two–Dimensional (2-D)

Space Science Vocabulary

Atmosphere
Air Resistance
Force
Mass
Gravity
Weightless
Satellite
Orbit
Diameter
Sphere
System

Evidence Circles About Air and Gravity

Evidence Circle Question #1: *Is there air on the Moon?*

1. **Show image #11 of the bootprint again.** Tell the students that this print, and others the astronauts left in the Moon's dust many years ago, probably look exactly the same today. They will probably still look the same millions of years from now.

2. **Discuss in evidence circles.** Ask students to discuss the question, "Is there air on the Moon?" in evidence circles. Assign them to teams of about four students, and remind them that each person in an evidence circle should have a chance to say what they think, and to back up their answer with evidence. The evidence can be from what they observe in this slide, other slides, video footage they have seen, or elsewhere.

3. **Share evidence with the whole class.** When teams are ready, ask a volunteer from each team to share their team's answer to the question, along with the evidence that team members used to back up their opinion.

4. **Evidence for "no air" that they may bring up.** Some evidence that your students may bring up to support the idea that there is no air on the Moon:

a. There was no air resistance to slow down the feather when it was dropped.

b. Sound could not travel on the Moon, except in space suits and spacecraft. Sound needs air to travel.

c. Astronauts breathing with air tanks in space suits is evidence that there was no air to breathe on the Moon.

d. Because there is no air, there is no wind, so the bootprint will probably last for millions of years. (The same is true for craters on the Moon: due to lack of wind and water, they don't erode, unlike craters on Earth.)

e. The flag needed a bar to keep it from hanging limp (again, no wind).

f. No parachute was used for landing on the Moon, because it wouldn't have worked without air.

Unit Goals

Earth, like all planets and stars, is round like a sphere.

Earth's gravity pulls things toward the center of Earth.

Gravity is everywhere.

TEACHER CONSIDERATIONS

TEACHING NOTES

Allow time for discussion: If some of your students are arguing that there is air on the Moon, be sure to spend enough time on this discussion to give them the opportunity to change their minds.

PROVIDING MORE EXPERIENCE

Optional Evidence Circle Question About Scale: Image #7, Earthrise, offers an opportunity to discuss size and distance as they relate to the concept of the scale of the Earth–Moon system and the connection between distance and apparent size of objects. If you have not presented Unit 1, *How Big and How Far?* you might consider posing the question below in an additional evidence circle. If your students struggle with the question, consider presenting the scale activities in Unit 1. If you have already presented Unit 1, you could use this question for review, either as an evidence circle or in a less formal manner:

Show Image #7, Earthrise. Ask, "If the Earth is really so much bigger than the Moon, why does the Moon look so much bigger than the Earth in this photo?"

What Some Teachers Said

"If my students could of spent the day looking at Apollo 11 pictures and exploring other Moon missions they would have loved it. The visuals make all the difference in the world! My 3rd graders were amazed that the footage was real."

"Lots of oohs & aahs"

Key Vocabulary

Science and Inquiry Vocabulary

Evidence
Scientific Explanation
Model
Scale Model
Prediction
Scientist
Three–Dimensional (3-D)
Two–Dimensional (2-D)

Space Science Vocabulary

Atmosphere
Air Resistance
Force
Mass
Gravity
Weightless
Satellite
Orbit
Diameter
Sphere
System

5. There is virtually no air on the Moon. Tell students that scientists have detected only the tiniest amount of air on the Moon. There is so little that it is generally said that there is no air on the Moon.

Evidence Circle Question #2: *Is there gravity on the Moon?*

1. Show Image #13, Astronaut with flag. Remind them that the flag got stuck as they were trying to stretch it out on the bar, like a curtain. Say that they will now discuss another question in their evidence circle.

2. Evidence circles. Ask students to discuss the question, "Is there gravity on the Moon?" in evidence circles. As before, they need to back up their answers with evidence from this image and other images, video footage they have seen, or elsewhere.

3. Share evidence with the whole class. After a few minutes, ask a few groups to share their evidence for gravity on the Moon. Some evidence that your students may bring up to support the idea that there is gravity on the Moon:

- The feather and hammer fell toward the surface of the Moon.
- The flag hung down from the bar because there is gravity on the Moon.
- Astronauts stepped, stood, ran, and jumped on the Moon, and they always came back to the ground.
- The astronauts made bootprints on the Moon. (The astronauts weren't floating: their weight pushed their shoes down into the dust.)
- Astronauts sank back into their seats as the spacecraft slowed down near the Moon.

Unit Goals

Earth, like all planets and stars, is round like a sphere.

Earth's gravity pulls things toward the center of Earth.

Gravity is everywhere.

TEACHER CONSIDERATIONS

OPTIONAL WRITING ASSIGNMENT AND ASSESSMENT OPPORTUNITY

After the evidence circle activity and the class discussion, consider having students individually write their responses to the two questions they discussed. This will provide an opportunity for students to recall and solidify what they have learned. Their written work can be assessed with the specific science concept rubric below or with the general rubrics provided on page 66.

Is there air on the Moon? Yes_____ No_____. List all the evidence you can think of to back up your answer:
Is there gravity on the Moon? Yes_____ No_____. List all the evidence you can think of to back up your answer:

	Understanding Science Concepts The key science concept for this assessment is the following: *There is virtually no air on the Moon.*	Understanding Science Concepts They key science concept for this assessment is the following: *There is gravity on the Moon.*
4	*The student demonstrates a complete understanding of the science concepts and uses scientific evidence to support the written explanation.* The student uses at least two or three pieces of evidence to explain that there is not air on the Moon. Evidence may include 1. There was no air resistance to slow down the feather when it was dropped. 2. Sound could not travel on the Moon except in space suits and spacecraft. Sound needs air to travel. 3. Since there is no air, there is no wind, so the boot print will probably last for millions of years. Also craters don't erode on the Moon since there is not wind and water. 4. The flag needed a bar to hold it up because there was no air to blow on it. 5. No parachute was used for landing on the Moon since it would not work without air.	*The student demonstrates a complete understanding of the science concepts and uses scientific evidence to support the written explanation.* The student uses at least two or three pieces of evidence to explain that there is gravity on the Moon. Evidence may include 1. The feather and the hammer fell towards the Moon. 2. The flag hung down from the bar because there is gravity on the Moon. 3. Astronauts stepped, stood, ran, and jumped on the Moon—when they did all of these things they came back to the ground. 4. Footprints on the Moon indicate that the weight of the astronauts pushed their shoes down into the dust. 5. Astronauts sank back into their seats as the spacecraft slowed down near the Moon.
3	*The student demonstrates a partial understanding of the science concepts.* Although understanding that there is no air on the Moon is demonstrated, the student does **not** tie the understanding together with multiple pieces of evidence from the class activities to create a cohesive explanation. They might use only one piece of evidence in their explanation. Or the quality of the "evidence" may be more generally experiential or rely on statements of presumed fact, rather than being drawn directly from classroom science experiences.	*The student demonstrates a partial understanding of the science concepts.* Although understanding that there is gravity on the Moon is demonstrated, the student does **not** tie the understanding together with multiple pieces of evidence from the class activities to create a cohesive explanation. They may use only one piece of evidence in their explanation. Or the quality of the "evidence" may be more generally experiential or rely on statements of presumed fact, rather than being drawn directly from classroom science experiences.
2	*The student demonstrates an insufficient understanding of the science concepts.* The student says that there is no air on the Moon, but the student does not use evidence from class to support the statement.	*The student demonstrates an insufficient understanding of the science concepts.* The student says that there is gravity on the Moon, but does not use evidence from class to support the statement.
1	*The content information is inaccurate.* Some possible inaccuracies are: 1. There is air on the Moon. 2. There is air on the Moon since the Moon has gravity. 3. There is air on the Moon since astronauts were able to go to the Moon.	*The content information is inaccurate.* Some possible inaccuracies are: 1. There is no gravity on the Moon. 2. There is no gravity on the Moon since there is no air on the Moon. 3. There is no gravity on the Moon since astronauts float on the Moon.
0	*The response is irrelevant or off topic.*	*The response is irrelevant or off topic.*
n/a	*The student has no opportunity to respond and has left the question blank*	*The student has no opportunity to respond and has left the question blank*

Quick Check for Understanding: Review question #4 and options A, C and E of question #2 on the *Pre-Unit 2 Questionnaire*. If there is still confusion, take a few minutes to discuss these questions now.

4. If a person is standing on the Moon holding a rock and then lets go of it, what will happen to the rock? It will __________

A. Float away.
B. Float where it is.
C. Fall to the ground.
D. Go up.

2. Which are true statements? Circle all that are true.

A. The Moon has no gravity because it has no air.
B. Gravity is an invisible force.
C. There is no gravity in space.
D. There is a pull of gravity between all objects.
E. If there were no air on Earth, people would float out into space.
F. Gravity keeps the Moon in Earth's orbit.

POST-UNIT 2 QUESTIONNAIRE, PAGE 1

Session 2.6 Student Sheet Name ______________________

Post–Unit 2 Questionnaire, Page 1

1. Which are true statements? Circle all that are true.

A. Gravity is an invisible force.
B. Gravity keeps the Moon in Earth's orbit.
C. There is no gravity in space.
D. There is a pull of gravity between all objects.
E. The Moon has no gravity because it has no air.
F. If there were no air on Earth, people would float out into space.

2. Why is the Earth flat in picture #1 and round in picture #2? (Circle the letter of the best answer.)

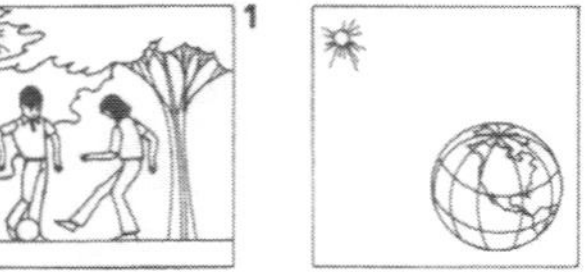

A. The Earth is round like a plate, so it seems round when you are over it and flat when you are on it.
B. The Earth is round like a ball, but it has flat spots on it.
C. The Earth is round like a ball, but it looks flat because we see only a small part of the ball.
D. The Earth is round like a ball, but people live on the flat part in the middle.
E. They are different Earths.

3. If a person is standing on the Moon holding a rock and let's go of it, what will happen to the rock? It will __________

A. Go up.
B. Fall to the ground.
C. Float where it is.
D. Float away.

(Over)

4. The Moon does have gravity, but less than Earth. Confirm that all the evidence tells us that there is gravity on the Moon. We also know that, like all other objects, the Moon has mass, so it does exert gravitational pull.

5. Gravity keeps the Moon in Earth's orbit. Tell students that the gravitational pull between the Earth and the Moon is what keeps the Moon in orbit around the Earth.

6. The gravitational pull of the Moon is six times *less* than the gravitational pull of Earth. The Moon's gravitational pull is strong enough to keep people from floating away, but weak enough so that people can jump higher there.

7. Post the last key concept. Say that many people think that you can't have gravity if there is no air. Ask what they might say to someone who said that. [It is Earth's gravitational pull that holds us on Earth, not the air. Places with no air, such as the Moon or other planets, still have gravity.]

Air is *not* necessary for gravity to exist.

8. There is confusion about air and gravity. Point out that many people are confused about this concept. They think that the Moon could not have gravity because it doesn't have air. Emphasize that any planet or moon has gravity, whether or not it has air. Gravity depends on mass. Gravity is everywhere.

Unit Goals

Earth, like all planets and stars, is round like a sphere.

Earth's gravity pulls things toward the center of Earth.

Gravity is everywhere.

OPTIONAL PROMPTS FOR WRITING OR DISCUSSION

Choose one or more of the prompts below for science journal writing in class or as homework. They could also be used for a discussion or during a final student sharing circle.

- Could you fly a kite on the Moon?
- Some things I've learned about gravity are:____________.
- Some things I've learned about the Moon are:____________.
- These are things I wonder about gravity and air:____________.
- Could there be a planet with no gravity?
- Can you ever escape gravity?

POST-UNIT 2 QUESTIONNAIRE, PAGE 2

Session 2.6 Student Sheet Name ____________________

Post–Unit 2 Questionnaire, Page 2

4. Pretend that the Earth is glass and you can look straight through it. Which way would you look, in a straight line, to see people in far-off countries such as China or India?

C. Upward
D. Downward
A. Westward
B. Eastward

A. Westward? **B.** Eastward? **C.** Upward? **D.** Downward?

5. This drawing shows some enlarged people dropping rocks at various places around the Earth. What happens to each rock? Draw a line showing the complete path of the rock from the person's hand to where it finally stops. Why will the rock fall that way?

____

____

____

____

____

____

____

____

____

PROVIDING MORE EXPERIENCE

Gravity on Other Planets in the Solar System: Many students have trouble understanding that gravity exists on other planets. Experience with calculating weights on other planets helps them grasp this idea. It also provides an opportunity for understanding that gravity is stronger with objects of greater mass. If you think that your students would benefit from more experience with gravity on other planets, you may choose to do one of the following activities with your students.

Calculating Your Weight on Another Planet: You could have students visit a website (such as http://kids.msfc.nasa.gov/Puzzles/Weight.asp or http://www.exploratorium.edu/ronh/weight/) where they can type in their own body weight, and find out how much they would weigh on other planets.

Continued on page 347

Taking the *Post-Unit 2 Questionnaire*

1. Find out how their ideas have changed. Tell your class that you think they have learned a lot since they took the Pre-Unit 2 Questionnaire, and that they probably know more about the Earth's shape and gravity than some adults do. Let them know that they will now get a chance to fill out the questionnaire again to see how their ideas have changed. Remind them that the ability to change one's ideas based on evidence is a sign of a good scientist.

2. Don't help one another. Remind them that the questionnaire is designed to find out what *each student* is thinking and to help you understand what they have learned.

3. Distribute questionnaires. You may want to tell students who finish early to read quietly or do other independent work while their classmates finish.

Unit Goals

Earth, like all planets and stars, is round like a sphere.

Earth's gravity pulls things toward the center of Earth.

Gravity is everywhere.

PROVIDING MORE EXPERIENCE, *CONTINUED*

Interactive Computer Game About Weight on Different Planets
In this simple Internet game, an object is weighed on a scale, and the weight is displayed. Off to the side is a chart of how much the object would weigh on different planets. The objective is to select the planet that the object is being weighed on by matching the weight of the object with the weight on the chart.
http://www.harcourtschool.com/activity/gravity/

Comparing Weight of an Object on Earth and the Moon
Gravity on the Moon is about $1/6$th as strong as gravity on Earth. To illustrate this, take two film canisters, or two other opaque containers identical to each other. Put six times as much weight in one as in the other. You can do this by adding and subtracting a heavy material such as sand, while weighing it. You can also do it by adding six washers, coins, or other semiheavy material to one container, and only one of the same object to the other container. Write "weight on Earth" on the heavier container, and "weight on Moon" on the lighter container. Allow students to feel the difference.

RESOURCES

The GEMS guide *Oobleck: What Do Scientists Do?* provides activities that focus on the Nature of Science and the connection between science and technology, especially as it relates to the history of space exploration. In this guide, students explore a mysterious substance said to have come from the surface of a fictitious planet, and discuss its strange properties in a scientific manner. Teams then design spacecraft capable of landing on the surface of the "planet." Later, students are shown the Mars Rover mission images.

Internet Resources on space for students:
An online kids' book on different outfits that astronauts wear inside and outside spacecraft:
http://www.nasa.gov/audience/forkids/home/F_Best_Dressed_Astronaut.html

An online kids' book on future missions to the Moon and Mars:
http://www.nasa.gov/audience/forkids/home/F_Vision_Slideshow.html

An online kids' book on the space shuttle:
http://www.nasa.gov/audience/forkids/artsstories/F_Space_Shuttle_is_Like.html

Key Vocabulary

Science and Inquiry Vocabulary

Evidence
Scientific Explanation
Model
Scale Model
Prediction
Scientist
Three–Dimensional (3-D)
Two–Dimensional (2-D)

Space Science Vocabulary

Atmosphere
Air Resistance
Force
Mass
Gravity
Weightless
Satellite
Orbit
Diameter
Sphere
System